Otniel Alencar Bandeira
Obede R. Alves
Lúcia Maria Moraes

Basic sanitation in Goiânia: A public health issue

Otniel Alencar Bandeira
Obede R. Alves
Lúcia Maria Moraes

Basic sanitation in Goiânia: A public health issue

Saneamento Básico em Goiânia: Uma questão de saúde pública

ScienciaScripts

Imprint

Cover image: www.ingimage.com

This book is a translation from the original published under ISBN 978-3-330-76730-0.

Publisher:
Sciencia Scripts
is a trademark of
Dodo Books Indian Ocean Ltd. and OmniScriptum S.R.L publishing group

120 High Road, East Finchley, London, N2 9ED, United Kingdom
Str. Armeneasca 28/1, office 1, Chisinau MD-2012, Republic of Moldova, Europe
Managing Directors: Ieva Konstantinova, Victoria Ursu
info@omniscriptum.com

Printed at: see last page
ISBN: 978-620-8-56665-4

ACKNOWLEDGEMENTS

To the Lord Jesus, for my life, strength, wisdom, inspiration and health. He alone is worthy of all honour and glory.

To my parents, Palmeri de Nazaré Coelho Bandeira and Maria Zelia Alencar Bandeira, for their love, affection and dedication, for always believing in me and supporting me in my endeavours. For having taught me to have faith and never give up on my dreams, for always striving for me to complete my undergraduate and master's degrees (...).

To my wife and research partner, Obede Rodrigues Alves, for her love, companionship, motivation and friendship. For never letting me give up in times of difficulty. For spending late nights awake by my side while I wrote this manuscript.

To my supervisor, Dr Lúcia Maria Moraes, for agreeing to supervise this work. Thank you for your patience and companionship.

To my younger brother, Palmeri Alencar Bandeira, for his patience and for always being willing to help when necessary.

To my brother, Estevão Alencar Bandeira, for his patience, trust and support during my master's programme.

PUC Goiás for the opportunity to do this master's degree.

To FAPEG for the scholarship.

To my beloved brothers and sisters in Christ from the Church in Goiânia and São Carlos, who have always been by my side, encouraging me and persevering in prayer.

To the professors of the Master's programme in Territorial Development and Planning.

To Dr Antônio Pasqualetto (PUC Goiás) and Dr Oyana Rodrigues dos Santos (IFG GOIÁS) for their collaboration during the preparation of the manuscript.

To my fellow Master's students.

SUMMARY

Sanitation is a set of measures adopted to preserve or alter the conditions of the environment in order to prevent disease, promote health, improve the population's quality of life, individual productivity and facilitate economic activity. Part of the Brazilian population lives in places where sanitation conditions are still very poor. Due to the lack of sanitation and minimal hygiene conditions, the population is subject to various types of illness such as diarrhoea, cholera, hepatitis A, dengue, chikungunya, zika, yellow fever, schistosomiasis, leptospirosis and others. This sparked an interest in analysing sanitation conditions in Goiânia, a regional metropolis of great economic importance to the state of Goiás. The following question arose: what is the current state of basic sanitation in Goiânia and what are the impacts caused by deficiencies in this service on the population's health? The importance of this work is justified by the relevance of sanitation for territorial planning in the municipality of Goiânia, with an emphasis on analysing the impacts on public health. In this sense, the aim of this research was to analyse the basic sanitation conditions of households in Goiânia and verify the impacts on the population's health due to diseases caused by the lack or inefficiency of these services. The research was divided into six stages, including a bibliographic survey of scientific databases, the Brazilian Institute of Geography and Statistics - IBGE and the SUS Information Technology Department - DATASUS, as well as the application of questionnaires with residents of Residencial Jardim do Cerrado I to IV. It was possible to see that sanitation services in Goiânia are still inefficient, especially with regard to the rate of sewage collection and treatment and the lack of rainwater drainage networks in most of the municipality. The number of hospitalisations of people with diseases related to the lack or precariousness of basic sanitation is also indicative of the population's sanitary conditions.

Keywords: Cities, diseases, urban planning, environmental health.

SUMMARY

INTRODUCTION

In recent decades, there have been many discussions about the concepts of urban and economic growth and development in Brazil and around the world. Alves and Belluzzo (2004) report that economic development should no longer be seen as a solution to problems such as human misery, i.e. growth alone is not a sufficient condition for promoting human well-being. In this respect, economic development has come to include social, environmental, cultural and political-institutional issues in an interconnected way.

According to the World Health Organisation (WHO) (2007), sanitation is defined as the control of all factors in the physical environment in which humans live that have or can have harmful effects on physical, mental and social well-being. Kassouf (1994) characterises sanitation as a set of socio-economic actions aimed at achieving environmental health in urban and rural areas, as discussed in this study.

Studies by the World Bank (1993) estimate that the inadequate domestic environment is responsible for almost 30 per cent of the occurrence of diseases in developing countries. Guimarães et al. (2007) point out that investing in sanitation is one of the ways to speed up a country's development and reduce the cost of hospitalisations in the public health system. Data released by the Ministry of Health (2010) shows that for every R$1.00 invested in basic sanitation, R$4.00 is saved in the area of curative medicine.

Since 2003, public sanitation policy in Brazil has been experiencing a new cycle represented by legal and regulatory frameworks, institutional restructuring and a resumption of investment. The creation of the Ministry of Cities (MCidades, 2005) and the National Secretariat for Environmental Sanitation (Secretaria Nacional de Saneamento Ambiental, 2004) has allowed government actions to be more focussed. With the holding of the Cities Conferences and the creation of the National Cities Council (ConCidades), it was possible to broaden the dialogue between the various organised segments of civil society and the state. Law 11.445/2007 ended a long period of uncertainty about the legal framework, inaugurating a new phase in the management of public basic sanitation services in the country, with planning taking centre stage in conducting and guiding public action. The resumption of investments at federal level, both with non-costly and costly resources, points to new strategies by the Brazilian state to tackle the deficits in sanitation services.

Part of the Brazilian population lives in places where sanitation conditions are still very poor. Due to the lack of sanitation and minimal hygiene conditions, the population is subject to various types of illness such as diarrhoea, cholera, hepatitis A, dengue, chikungunya, Zika, yellow fever, schistosomiasis, leptospirosis and others. The literature on public health shows that the lack of drinking water and sanitation is one of the main causes of child mortality. In 2010, 20 children between the ages of 0 and 5 died every day in Brazil as a result of a lack of drinking water and, above all, sanitary sewage. In other words, every 72 minutes a child between the ages of 0 and 5 dies due to a lack of basic sanitation services (MINISTÉRIO DA SAÚDE, 2010; LISBOA et al., 2013), According to Amaral (et al, 2003), water-borne diseases are mainly caused by pathogenic microorganisms that have their origins in the socio-economic factors of the population living in precarious conditions in favelas, on stilts, etc. There are still many regional inequalities in Brazil, and this is determined by the capitalist and dichotomous precepts that govern society. As one of the consequences of the economic and social inequalities between states and municipalities, there is a lack of access to infrastructure and quality of basic sanitation services, as shown in the Jardim do Cerrado housing estate in Goiânia. In places where basic sanitation services do not exist, or are inefficient, they are prone to the occurrence of the different diseases already mentioned and consequently cause countless inconveniences to the health and well-being of residents and also to the social and economic development of the region.

PROBLEM

Considering basic sanitation, a public health issue in the city of Goiânia, the guiding axis of this research, the following question arises: what is the current condition of basic sanitation in households in Goiânia and what impacts do deficiencies in this service have on the population's health? How does Residencial Jardim do Cerrado, a neighbourhood planned in the 21st century, still promote differentiated places with social inequality for its inhabitants?

HYPOTHESIS

The following research hypotheses were defined: 1) There are a significant number of households in Goiânia with basic sanitation deficiencies, especially in relation to sewage; 2) The number of cases of illness, hospital admissions and deaths caused by the lack or inefficiency of basic sanitation in Goiânia is still high, especially in the first years of life, the lack of sewage systems in Jardim do Cerrado in stages I, II, III and IV may be a reason associated with the government's neglect of the low-income population.

BACKGROUND

In view of the deficiencies in sanitation and the impacts on the population's health, it is necessary to intervene in defence of the environment, promotion of public health and improvement of sanitary conditions, with special emphasis on urban areas, where the Brazilian population is predominantly concentrated. Against this backdrop, this research is justified because it deals with the relevance of sanitation for territorial planning in the municipality of Goiânia, with an emphasis on analysing the impacts on public health.

OBJECTIVES

The sustainable planning of cities, especially Goiânia - the subject of this research - occurs when the availability and efficiency of basic sanitation, which must be universalised, is taken into account. With this in mind, the following objectives were set:

1. To analyse the basic sanitation conditions of households in the city of Goiânia;
2. To survey the impact on the health of the population of Goiânia, due to

the diseases caused by the absence/inefficiency of basic sanitation services;

3. To assess the basic sanitation conditions in Residencial Jardim do Cerrado I to IV and to verify the residents' opinion of the sanitation services offered in the region.

METHODOLOGY

The research was divided into five stages: 1) Bibliographic survey in scientific databases; 2) Survey of data on sanitation conditions of households and residents of Goiânia from the IBGE demographic censuses; 3) Survey of data in the Department of Informatics of SUS - DATASUS, on specific diseases related to deficiencies in basic sanitation; 4) Collection of primary information on basic sanitation conditions in Jardim do Cerrado Residences I to IV; 5) Application of questionnaires with residents of Jardim do Cerrado Residences I to IV; 6) Tabulation, preparation of tables and graphs and subsequent analysis and interpretation of the data collected.

In the theoretical field, the bibliographic survey was carried out by searching scientific databases such as Scielo, Portal Periódico CAPES, Science Direct, Scopus and the Virtual Health Library (BVS). Next, texts relevant to the subject were selected, going through the three reading processes defined by Gil (1994), namely: exploratory reading (to see if there was interest in the research), selective reading (to determine what would be used in the work) and finally, analytical reading (to determine their order). Secondary data was extracted from the IBGE Demographic Censuses (1991, 2000 and 2010), DATASUS (2008 - 2016) and the Goiânia Municipal Health Department (2014-2016).

According to Selltiz et al. (1975) and Gil (2002), research can have an exploratory or descriptive and causal focus, with exploratory studies aiming to get to know or understand the element, and their basic characteristic being the non-existence of hypotheses. Causal or descriptive studies, on the other hand, seek to investigate plausible cause and effect relationships, and the existence of previous hypotheses and the identification of factors that contribute to the occurrence of phenomena are essential conditions for the success of the research.

This research is classified as descriptive, since it seeks to expose the condition of basic sanitation services in the households and residents of Goiânia, based on the IBGE Demographic Census, and subsequently present an analysis of hospital admissions to the Unified Health System (SUS), public spending and deaths due to diseases related to the absence or inefficiency of basic sanitation services. To this end, four groups of diseases were selected, namely:

1) Faeco-oral transmission diseases (amoebiasis, diarrhoea, cholera, typhoid and paratyphoid fevers, polio sequelae and viral hepatitis - except type B);

2) Diseases transmitted by insect vectors (Dengue virus haemorrhagic fever, Classic Dengue, Chikungunya, Zika, Yellow fever, Filariasis, Leishmaniasis);

3) Diseases transmitted by contact with water (Schistosomiasis and Leptospirosis);

4) Diseases hygiene-related diseases: Eye diseases (trachoma and conjunctivitis) and skin diseases (mycoses)

In order to better understand the data, it was separated into four universes of study, namely: 1) number of hospital admissions by disease, 2) total amount spent on admissions, 3) admissions by age group and 4) number of deaths by disease. Information on diseases was obtained from the SUS Information Technology Department (DATASUS) from January 2008 to November 2016. This period was chosen due to the existence of information in DATASUS for the four study universes adopted for this research. The information on dengue fever, chikungunya fever and Zika was obtained from the Goiânia Municipal Health Department, and the periods studied for the last two diseases differ from the others because they have only recently been notified in Brazil. Thus, for chikungunya fever, the evaluation period was from 2014 to September 2016 and for Zika it was from September 2015 to September 2016.

CHAPTER STRUCTURE

The manuscript is structured in four chapters, in addition to this introductory section, considerations and references. The first chapter deals with questions about planning, a brief history and the legal instruments for urban and environmental planning in Brazil. In recent decades, concerns have been raised about the quality of life in cities, given that population densification, without proper planning, generates a series of negative consequences for urban life, including basic sanitation.

The second chapter deals with the relationship between basic sanitation, health and quality of life. It presents a brief contextualisation of sanitation issues and the sector's main legal framework, from the time of colonial Brazil to the present day. It provides an overview of the basic sanitation situation in Brazil by macro-region and population served by drinking water supply, sewerage, solid waste management and urban drainage services. From this overview it is possible to see the profound regional inequalities that exist in the country, one of the main challenges to universalising access to these services.

The third chapter is divided into two parts: the first part presents an overview of the basic sanitation conditions of households in Goiânia with regard to water supply, sewage disposal and solid waste management based on the IBGE censuses of 1991, 2000 and 2010. The second part presents data on hospital admissions to the SUS for diseases related to poor or inefficient sanitation in Goiânia, as well as information on mortality and public spending on hospitalisation services.

The fourth chapter presents a characterisation of the Jardim Cerrado Residences and some comparisons between the modules intended for the poorest families, belonging to band one of the Minha Casa Minha Vida Programme (PMCMV) and the modules intended for bands two and three of the programme. Data is also presented on the survey of basic sanitation conditions in Residencial Jardim Cerrado I to IV, in the four areas of sanitation: water supply, sewage, solid waste and rainwater drainage. This chapter also presents the answers given by residents when applying the questionnaire about the sanitation services offered in the neighbourhood.

CHAPTER 1

URBAN AND ENVIRONMENTAL PLANNING IN BRAZIL

1.1 Urban planning in Brazil

According to Lafer (1973), planning has instruments that point the way to its full development. These include tactics, which deal with short- and medium-term challenges, and objectives, which deal with long-term issues. Planning can be applied to both administrative and urban areas.

From an administrative point of view, strategic planning, according to Cavenaghi (2009), is a possible action for any enterprise, through which it aims to obtain competitive advantages, based on a good strategy aimed at positive solutions. For Cabanillas (2005), strategic planning represents the definition of goals and objectives that can provide good results.

From an urban perspective, Villaça (1998) says that planning is a way of proposing structural changes in order to create a better future for the city and its inhabitants. In city planning, long-, medium- and short-term goals are set and benefits are identified that enable a holistic view of the urban reality. In this vein, Souza (2008) emphasises that actions are included through popular participation and by the municipality's technical staff, who are linked to urban planning. It is through planning that new plans and programmes emerge to promote the country's development.

The social approach to planning introduced by Castells (1983), in his classic work "The Urban Question", considers urban space as a product of a given social formation and therefore states that:

> Urban planning is political intervention in the specific articulation of the different instances of a social formation at the heart of a collective unit for the reproduction of the labour force, with the aim of ensuring its expanded reproduction, regulating non-antagonistic contradictions, thus ensuring social class interests in the social formation as a whole and the reorganisation of the urban system in order to guarantee the structural reproduction of the dominant mode of production . (CASTELLS, 1983, p. 376-377).

In Villaça's (1998) words, planning urban space means looking to the future of cities, seeking precautionary measures against problems and difficulties, or even making better use of possible benefits that should be applied to *cities.*

These concepts reinforce Maricato's (2008) idea that cities undergo constant changes over time, because their population is in constant interaction with the environment and space, altering them to meet their needs. Public managers must therefore pay attention to these changes and apply urban planning and environmental instruments in order to reduce social inequalities and promote a more balanced development of the city.

In the 1990s, urban public policies in Brazil sought to help reduce social inequalities in the urban environment and promote debates between different social agents to establish the main guidelines for masterplans. During this period, it emerged that Brazilian cities shared similar urban problems, such as conurbation[1] , accelerated expansion of the peripheries, pressure from the property sector for changes in land use and occupation, among others (VILLAÇA, 2000).

One of the limiting factors that makes it difficult to carry out city planning efficiently is the gap between what is planned and established in masterplans and the city's production process, as shown by Costa (et al., 2012). The author emphasises that the most important thing is not just the growth of the city, but the way in which it grows and develops. Managers and society in general must pay attention to this, otherwise the city's urban and social problems will take a direction that will be difficult to reverse or solve in the future.

Urban planning in Brazil has been developed through various modalities, including zoning, beautification plans, improvement plans and, in recent decades, the masterplan. Other modalities mentioned by Villaça (1999) are physical territorial planning, planning for new cities such as the city of Goiânia, control of land use and occupation and sectoral planning. Historically, zoning was one of the first tools used in urban planning in Brazilian cities, as was the case in Rio de Janeiro and São Paulo in the mid-19th century, albeit in a very rudimentary way.

According to Villaça (2000), urban planning experiences in Brazil have gone

---

[1] It's a term used to describe an urban phenomenon that occurs when two or more cities/municipalities come together to form a single urban network, as if it were just one city.

through different periods with different visions and practices. In Brazil's current urban planning legal framework, the prevailing concept is that a city needs to have a Master Plan in order to have any level of urban planning and that the lack of such a plan is closely related to high crime rates due to the lack of urban/community facilities, services and quality infrastructure. Therefore, the prevailing idea is that in order to achieve orderly cities, equipped with infrastructure and low rates of urban violence, planning, management and effective plans are necessary

Analysing the trajectory of urban planning in Brazilian cities, there are few moments when public investments and works in the city go in the same direction. On the one hand, autonomous public and private actions and interventions predominate, unrelated to the planning and plans in force, while on the other hand there are planning and plans that are not reflected in works and improvements within the territory of the cities. In this way, Cymbalista (2006a) emphasises that Brazilian cities have not achieved orderly, socially inclusive and environmentally sustainable growth because planning without action is as ineffective as action without planning.

No matter how good the technical quality of the plans, they alone are not capable of solving structural urban problems in Brazilian cities such as flooding, occupation of risk areas, congestion, the absence/inefficiency of public transport, environmental sanitation, decent housing for all and the urbanisation of working class neighbourhoods. Feldman (1996) emphasises that urban problems increase as there is no administrative discontinuity and the erratic nature of urban policies leads to ineffective plans and consequently to the disorganisation of cities. Lack of investment is a factor that influences city disorder and can be more serious than non-compliance with a plan.

The development process is an integrated process in the various aspects of a reality. The assumption that social development is a consequence of economic development has led to unsatisfactory results, since setting only economic goals results in transformations limited to the purely economic level, leaving the social structure and consequently the social problems unchanged (COSTA et al., 2012). The history of Brazilian planning is systematised by Villaça (1999, p. 169 - 244) into five main currents, as shown in Figure 1.

1. Santtarist urbanism	With the aim of hygienising cities, it was introduced at the beginning of the 20th century.
2. Planning new cities	It guided the emergence of new city projects, such as Belo Horizonte. Brasilia. Palmas, etc.
3. zoning	It is considered the oldest urban planning instrument in Brazil.
4. Urban infrastructure plan	Infrastructure projects/design considered.
5. Urban planning Scrictu Sensu	Object of current urban planning in Brazil

Figure 1 - Five main currents of urban planning in Brazil Source: Villaça, 1999, p. 169.

These five currents occur in periods that can be categorised as follows: Urban planning, from the 19th century to the 1930s, from the 1930s to the 1990s and from the 1990s to the present day. In the corresponding period, between the 19th century and the early 1930s, zoning, beautification and improvement plans, "sanitary" urbanism and infrastructure plans were active. In the subsequent period, from the 1930s to the early 1990s, the currents of urban infrastructure plans, *strictu sensu* planning and the planning of new cities were active, in particular the 1933 city of Goiânia, the subject of this research, and the 1957 Brasilia Plan.

1.2 0 Environmental Planning in Brazil

Post-modern urban planning, according to Silva et al. (2013), is increasingly linked to the environmental process and the respective legal instruments, demanding multidisciplinary knowledge from professionals in the field, as well as new forms, methods and applications of concepts that tend to accompany the dynamism of today's society. Rodrigues et al. (2011) show that with the intense process of urbanisation, the city has come to exercise more forceful economic and social functions, configuring itself as a built environment.

One of the main objectives of urban planning and development is to provide an *upgrade* in society's living conditions, because according to the Federal Constitution of 1988 in article 225:

> Art. 225. Everyone has the right to an ecologically balanced environment, a good for the common use of the people and essential to a healthy quality of life, and the public authorities and the community have the duty to defend and preserve it for present and future generations (BRASIL, 1988, Article 225).

The City Statute, Law No. 10,257 of 2001, reinforces this view in Article 1°, sole paragraph:

> Art. 1 Sole Paragraph. For all intents and purposes, this law, known as the City Statute, establishes rules of public order and social interest that regulate the use of urban property in favour of the collective good, the safety and well-being of citizens,

as well as environmental balance (BRASIL, 2001, article, 1, sole paragraph).

According to the World Health Organisation (WHO), quality of life is directly related to certain indicators such as food conditions, income, education, sanitation, health and others. These indicators are influenced by public policy and public managers (DI SARNO, 2004). Dias (2005) emphasises that the environmental issue is directly related to the population's quality of life and must be taken into account by city planners.

In the last three decades, environmental planning has emerged due to the dramatic increase in competition for land, water, energy and biological resources, which has generated the need to organise land use, make this use compatible with the protection of threatened environments and improve people's quality of life. Environmental planning came about as a solution to recurring conflicts between the goals of environmental conservation and technological development (ANTONUCCI et al., 2010).

Almeida et al. (1999) describes that environmental planning needs to be seen with a holistic vision, because the decision-making and interaction processes related to the environment are complex and involve a diversity of human activities. According to Albano (3013), environmental planning is an essential element for socio-economic development aimed at making the best use of a territorial space by identifying its potential and weaknesses. This topic is addressed in this research due to the precariousness of basic sanitation in Brazilian cities, especially in the city of Goiânia.

Environmental planning makes it possible for cities to develop in a more sustainable way, as shown by researchers Canepa (2007) and Franco (2001), i.e. in harmony with economic, political, social, cultural and environmental issues. This idea is reinforced by Maria (2013) and Albano (2013), who emphasise that one of the great challenges of environmental planning is to be able to link the natural environment to interurban processes, through the association between planning and environmental analysis, rational use of natural resources and a better quality of life for the population.

In the words of Dias (2005), it is important to emphasise that there is no sustainable development standard applicable to all cities, which is why it is the responsibility of each management body to survey the real situation of the municipality in order to establish goals, strategies and solutions to urban problems and achieve good quality of life indicators, as will be shown later in this study.

In the 1980s, the term environmental planning was understood by many to mean the planning of a region, according to Maria (2013 p. 27) "the planning of a locality with a view to integrating information and research into the environment, which provides for actions and standards along ethical lines of development". In this respect, there are those who are concerned with the packaging of utensils and the consequent impacts of the socio-economic logics that determine a place of interest. In this way, the author states that: "the principles of environmental planning relate directly to the term sustainability and interdisciplinarity, which, as far as it is concerned, requires a holistic view of analysis in order to be applied". In a simple way, she emphasises that planning should be carried out in a triad, environment-human-society, which is now seen as a single unit (MARIA, 2013).

Environmental planning consists of communicating and integrating the principles that make up the environment, as Albano (2013) points out when he writes: "the function of determining the relationships between ecological measures and community processes, socio-cultural needs, economic activities and interests" are essentially fundamental, and must have the intention of maintaining the concept of the possible integrity of its elements and components. Planning from this point of view is generally systematic and holistic, but it has a "process of first identifying the space and then integrating it", as explained by Albamo.

Environmental planning has a technique for establishing actions within a context and not in particular. Antonucci et al. (2010) have worked with the concept of resource, which recognises the natural element as a source of material for man. It envisages the actions of many sectors of society, through their representatives, with society having the right and duty to think about the issues that concern it. Three axes need to be determined: technical, social and political.

According to Maria (2013, p. 24), several planners at the present time talk about the "objective of enhancing the quality of life of human beings, followed by maintaining the processes of nature and its heterogeneity". Others argue that the basis of environmental planning should be decentralised, with the participation of the population *in loco*, with "multiple interlocutors and maximum participatory activity, with the possibility of introducing popular councils".

In short, for Albano (2013), environmental planning, at least in Brazil, does not efficiently describe reality, nor does it achieve the ideals it advocates. The time has come to reflect on the efficiency of theoretical discourse, as well as on the construction of theory and method. These are currently the major obstacles and the greatest challenges for this area of knowledge (ALBANO, 2013).

1.2.1 Environmental legislation

Brazilian environmental legislation has rules and principles based on the 1988 Federal Constitution (CF), especially the National Environmental Policy (PNMA) Law No. 6.938/81. The issue is dealt with in a special way in the Federal Constitution, Article 225 of which makes it clear that the environment is a good for the common use of the people, i.e. it is an invaluable social asset that cannot be individualised. Siqueira (2002) emphasises that these riches can be tangible (such as forests, rivers, fauna) or intangible (such as the history of a community, culture, religion, rituals, typical foods).

In Brazil, environmental and naturalist documents can be found as far back as the time of the Empire, in the first decades of the 1800s, when problems linked to the impact of human activities on natural resources were discussed. "The documents written by D. João VI and D. Pedro II that guided the first environmental protection regulations were written by naturalists brought to Brazil by the Empire", such as Martius, Mikan, Pohl, Spix, Natterer and Loefgren, who were mainly concerned with the protection of forests, "the quality and availability of water resources and the sanitation of cities" (FELDMAN, 1996).

The Forest Code and the Water Code emerged in the 1930s. The first Forest Code, in 1934, came about because of the boom in coffee plantations. As the plantations expanded, the forests began to become very distant, which made transport difficult and the cost of firewood high. Silva Filho (et al. 2015) adds that this was followed by the Forest Code, which obliged landowners to maintain 25% of the area with original forest cover. During this period there was no study or guidance as to which area should be preserved.

Silvestre's (2008) research shows that in the same year the Water Code came into being, the result of a set of initiatives proposed since the 1930 Revolution, which aimed to boost Brazil's growth by making it a modern, industrialised and developed country.

During the 1950s and 1970s, while Brazil was going through a process of industrialisation, environmental issues were very much trivialised and governments cared little about the pollution of natural resources. From the 1960s onwards, environmental concern grew in the United States, largely among the civilian population and non-governmental organisations, driven by the emergence of the ecological movement, the so-called North American environmental revolution, which began after the publication of Rachel Carson's book[2] (MONTIBELLER-FILHO, 2008). After a few years, Canada, Japan, New Zealand, Australia and Western Europe also joined the environmental debate.

The concept of sustainable development emerged under the name of eco-development in the 1970s and was consolidated with the Stockholm and Rio 92 Conferences (ROMEIRO, 2012). The Stockholm Conference in 1972 and the Rio de Janeiro Conference in 1992 were two key events organised by the United Nations (UN) to guide environmental issues at international level. The aim of the first event was to raise international awareness of the importance of maintaining air quality in large urban centres, and of maintaining the cleanliness and quality of water resources.

Since then, the preservation of natural resources has been accepted by various countries and the environmental issue has become a global concern and is now on the agenda of political discussions and international negotiations (BRASIL, 2007; SILVA, 2011).

The main aim of the Rio Conference in 1992 was to discuss the proposals of

---

[2] Rachel Carson, author of the book "Silent Spring", which deals with the environmental and public health risks caused by the indiscriminate use of pesticides.

the report "Our Common Future", produced in 1987 by the World Commission on the Environment, a commission set up by the UN at the end of 1983 on the initiative of UNEP[3] (BASSANI and CARPIGIANI, 2010). This report sought to establish a balance between development and the preservation of natural resources, thus highlighting the concept of sustainable development, defined as "that which meets the needs of the present without compromising the ability of future generations to meet their own needs" (BRASIL, 2007, p. 13).

Before Rio 92, Brazil had already entered into these environmental debates, after coming under a lot of pressure from international banks, which began to demand environmental impact studies for project financing, a fact that forced Brazil to create the National Environmental Policy in 1981 (COSTA et al., 2012).

With the creation of Law 6.938 of 1981 - which instituted the National Environmental Policy (PNMA), created the National Environmental System (SISNAMA) and the National Environmental Council (CONAMA) - there was the integration of bodies and institutions from all spheres of government involved with the environmental issue and the expansion of discussions on the subject to various segments of society (SAULE JÚNIOR, 2004). One of the objectives of the PNMA was to make economic and social development compatible with the preservation of environmental quality and ecological balance and the preservation of environmental resources (BRASIL, 1981). Before the PNMA, legal guidelines were sectorised, linked to one aspect of the environment such as forest preservation, fauna protection, water resource conservation or pollutants (Chart 1).

Table 1 - Environmental Legislation: Main Legal Documents

TYPE OF STANDARD	DATE	SUBJECT
Decree 24.643	10.07.1934	Establishes the Water Code.
Law No. 5,197	03.01.1967	Provides for the Protection of Fauna.
Decree Law No. 221	28.02.1967	Provides for the protection and encouragement of fishing.
Law No. 6.513	20.12.1977	Provides for the creation of Special Areas and Sites of Tourist Interest.
Law No. 6.938	31.08.1981	Provides for the National Environmental Policy.
CONAMA Res. No. 001	23.03.1986	Establishes guidelines for environmental impact assessment.
Law No. 7.511	07.07.1986	Amends provisions of Law 4.771, of 15 September 1965, which establishes the new Forest Code.

[3] The United Nations Environment Programme (UNEP) was created in 1972 with headquarters in Nairobi, Kenya (BRAZIL, 2007).

CF Brazil	05.10.1988	Chapter VI - The Environment: Article 225.
Law No. 7.804	18.07.1989	Amends Law No. 6938, of 31 March 1981, which provides for the National Environmental Policy.
Decree 99.274	06.06.1990	Provides for ecological stations.
Decree No. 1.354	29.12.1994	Establishes, within the scope of the Ministry of the Environment, Water Resources and the Legal Amazon.
Law No. 9.433	08.01.1997	Institutes the National Water Resources Policy.
Law No. 9.605	12.02.1998	Environmental Crimes Law.
Law No. 9.795	27.04.1999	Provides for Environmental Education.
Law No. 9.985	18.06.2000	Establishes the National System of Nature Conservation Units . Regulates art. 225, §1, items I, II, III and VI of the Federal Constitution.
CONAMA Res. No. 302	20.03.2002	Provides for the parameters, definitions and limits of Permanent Preservation Areas.
CONAMA Res. No. 303	20.03.2002	Provides for the parameters, definitions and limits of Permanent Preservation Areas.
Decree No. 4703	21.05.2003	Provides for the National Biological Diversity Programme.
Law No. 11.105	24.03.2005	It has established inspection systems for the various activities involving genetically modified organisms.
Law No. 11.428	22.12.2006	Provides for the utilisation and protection of the native vegetation of the Atlantic Forest Biome.
Decree No. 6.288	06.12.2007	Consolidates criteria for the EEZ.
Law No. 12.305	02.08.2010	Institutes the National Solid Waste Policy.
Law No. 12.651 (Forest Code)	25.05.2012	Provides for the protection of native vegetation.

Source: Prepared by the author based on data from SANTOS (2013).

In 1986, another extremely important legal document was approved: CONAMA Resolution 001/86, which made environmental impact studies mandatory in Brazil for a wide range of human activities. This resolution established definitions, responsibilities, basic criteria and general guidelines for the use and implementation of Environmental Impact Assessment (BRASIL, 1986). The resolution not only defined environmental impact, but also described the undertakings required to submit an Environmental Impact Assessment (EIA) and created Environmental Impact Reports, as an expression of the results of the EIA, but in simpler language accessible to the community in general. Subsequently, CONAMA Resolution 06/1987 established rules for the environmental licensing of large-scale works related to electricity generation, establishing the prior licensing of these activities (COSTA et al., 2012).

The definition of environmental impact provided by CONAMA Resolution No. 001 of 1986 is described in Article 1°: "Environmental impact is considered to be any alteration of the physical, chemical and biological properties of the environment, caused by any form of matter or energy resulting from human activities" (...) (BRASIL,

1986). Sánchez (2006, p. 30) points out that this definition is more related to the concept of pollution and not to environmental impact, due to the fact that it mentions "any form of matter or energy" as a determining factor by "altering the physical, chemical or biological properties of the environment". The author defines environmental impact as "alteration of a natural or social process resulting from human action".

Before Law 6.938/81, the Brazilian government controlled the use of natural resources through the Water and Mining Code and the first Forestry Code, Law No. 4771/1965. As far as the urban environment is concerned, an important advance was made with the 1965 Code, which defined preservation areas, with the enactment of the Land Statute Law No. 6.453/1964, which created the conditions for public authorities to interfere in economic activities that transform the environment, thus enabling the emergence of new environmental laws (ROLNIK, 2003). The Brazilian Forest Code underwent several changes and came into force through Law No. 12.651 of 2012. It can be said that the main reason for amending the 1965 forestry code was to reconcile the interests of agriculture with the protection of natural resources (LIMA et. al., 2010).

The greatest difficulties in applying the current forestry code relate to the restoration of permanent preservation areas (APP's) and the recovery or compensation of the legal reserve of properties that will take place through the Environmental Regularisation Programme (PRA), registration in the Rural Environmental Registry (CAR) and the signing of a Term of Commitment (TC), given the difficulties in applying these instruments, the lack of inspection, doubts regarding the quality of the information and the slowness in implementing the system (METZGER, 2010).

1.2.2 Environmental Planning Structure and Instruments

In the 1990s, as Cymbalista (2006) shows, environmental planning was incorporated into municipal master plans. Despite the progress made, the interests of the economy are still predominant in the planning decision-making process, and there are still major challenges to overcome in order to build planning based on environmental conservation and social quality.

Environmental planning is organised within a process that involves research, analysis and synthesis. Research aims to gather and organise data for a better understanding. The organised data is evaluated to understand the study, with its successes and failures, constituting the analysis phase. Synthesis refers to the application of the knowledge gained for decision-making. And to fulfil these stages, planning is generally presented as a system, developed in phases that evolve gradually, the result of one is the basis or principles for the development of the next phase (MARIA, 2013).

The frequent stages in environmental planning are defining objectives, diagnosing, analysing alternatives and making decisions. But in practice, it's not as simple as that. According to Souza (2008), this is because there are different conceptions of environmental planning, different objects and various methodological structures for drawing up and implementing projects.

According to Rodriguez (1991), environmental planning consists of five phases:

> methodological and operational implementation; analysis and systematisation of environmental indicators; diagnosis of the environment with identification of impacts, risks and efficiency of use; preparation of a territorial organisation model; proposal of measures and implementation of management mechanisms (RODRIGUEZ, 1991, p. 15).

Santos (1998), on the other hand, presents a planning process divided into eight phases: definition of objectives, definition of the organisational structure, diagnosis, evaluation of successes and conflicts, integration and classification of information, identification of alternatives, selection of alternatives and decision-making, guidelines and monitoring. In this proposal, the eighth phase refers to public opinion.

In order to achieve these environmental quality indicators, environmental planning has a number of instruments at its disposal: Zoning (environmental, ecological-economic); Environmental Impact Assessments (EIA); River Basin Plans; Water Resources Management Plans; Environmental Licensing; Solid Waste Management Plans; Sanitation Plans; Management Plans or Environmental

Protection Areas (APA); Environmental Master Plans; Risk Maps; Environmental Education; Grants and Concessions and others.

Zoning is made up of two distinct phases: inventory and diagnosis. Through these phases, areas that share similar environmental and urban spaces are defined. Generally, the planning unit for a zone is the river basin. It should be noted that in Brazil, zoning is used by the public authorities as a legal instrument, regulating the use of the national territory (SILVA, 2003).

The City Statute (BRASIL, 2001), which standardised articles 182 and 183 of the 1988 Federal Constitution and deals with urban policy, aims to prioritise the social function of the city and urban property. However, this responsibility to intervene in the territory was delegated to the municipalities, which had to draw up and approve the master plans that became urban planning instruments of great value for the urban policy of any city with more than 20,000 inhabitants, as Rolnik (2001) explains.

The Masterplan is a basic mechanism for regulating development policy and guaranteeing quality of life in the municipality. Santos (2004) highlights the fact that it applies and creates planning instruments in conjunction with society in order to harmonise land use and occupation, economic processes and the city's infrastructure. According to the same author, the Master Plan assumes a relevant role when it identifies the aspirations of the community and creates means to guarantee and encourage popular participation in municipal management processes.

In Fidalgo's (2003) view, the Masterplan should be considered a planning instrument when it aims to improve relations between man and nature, when it has clear and well-consolidated political objectives and goals through its proposed guidelines and actions, and when it draws up a diagnosis that is concerned with natural resources and man.

It is also important to note that when urbanisation takes place on natural systems, without proper planning and management, with mitigation of failures, environmental malpractice occurs, reducing the capacity of cities to implement sustainability. Consequently, sustainable urbanisation "are cities that implement sustainable public policies that respect and preserve the environment" is a characteristic that is growing rapidly (CARRERA, 2005 apud CANEPA, 2007).

Environmental licensing is the administrative procedure by which the competent environmental body licenses the location, installation, operation and expansion and/or renewal of undertakings and activities that use environmental resources, and is regulated by Laws No. 6.938/81, 9605/98, Decree 3.179/99 and 99.274/91, as well as various resolutions of the National Environment Council (CONAMA).

The environmental impact study is an impact assessment tool that systematically demonstrates the consequences of implementing a project on the environment. The study points out the environmental, economic and social impacts of the project in a given location, as well as compensatory, mitigating and environmental monitoring measures. In addition to the EIA, there are also Neighbourhood Impact Studies (EIV), Environmental Control and Management Plans, Traffic Impact Studies (EIT) and others. Sánchez (2006, p. 161) states that the EIA "is the most important document in the entire environmental impact assessment process. It is on this basis that the main decisions regarding the environmental viability of a project will be made".

Water resource plans are more comprehensive and use similar structures to environmental planning. For these lessons, the river basin is the territorial space on which planners agree. From this point of view, many names have been given to these studies, such as environmental water resources planning, river basin planning (PBH), water resources planning, river basin management, or masterplan for river basin management (SANTOS dos, 2013).

It should be noted that each name is given a concept, linked in particular to the objective, aspect and expected action of the process. It is therefore inevitable to reflect on whether planning makes it possible to manage the resource, order the space, carry out tasks, manipulate the environment, suggest alternatives, implement projects, monitor, control events, exploit resources associated with water or supply population centres, among other actions. Planning should promote and guarantee the protection of natural criteria, but not all of this is done, leading to many misunderstandings (ALMEIDA et al., 1999).

The preservation of biodiversity refers to the resolution of conflicts, which are important and precede the definition of planning procedures. Territorial systematisation through zoning and the establishment of action programmes in the form of environmental regulations are essential parts of the plan. Planning is usually harmonised and conceived within a systemic, integrated and frequent analysis, with proposals for a horizon of a few years.

A crucial expression that planners should note is that, whatever environmental planning tool is established, it always works with a cut-out of the reality of the space and, consequently, the complexity and relationships of the environment are simplified and generalised. For Albano (2013), the best performance lies in identifying comprehensive and tangible objectives of the variables that most faithfully affect the main existing relationships and fundamental difficulties in the real and future scenario of the planned space.

CHAPTER 2

BASIC SANITATION IN BRAZIL

Sanitation is a set of measures adopted to preserve or alter the conditions of the environment in order to prevent disease, promote health, improve the quality of life of the population and individual productivity, as well as facilitating economic activity (HELER, 1998). The World Health Organisation (WHO) defines sanitation as the control of factors in the physical environment that can cause harmful effects on the physical, mental and social well-being of a population (PHILIPPI JR; MALHEIROS, 2005). It can also be said that sanitation is the set of socio-economic actions aimed at achieving environmental health[4] and providing better living conditions for the population.

Basic sanitation is one of the main indicators that makes a country developed, because it directly affects the social, environmental and economic aspects of a territory. The provision of basic services such as treated water, sewage and solid waste collection and treatment, provides a better quality of life for people, reduces infant mortality, has a positive impact on students' school performance, maximises the exploitation of a region's tourist potential, increases the value of real estate, indicates greater worker productivity and consequently income, reduces health costs, etc.

It's important to note that in the state of Goiás in the 1930s, Pedro Ludovico used Vila Boa's physical, health, environmental and social conditions as one of his justifications for transferring the capital to the region of Campinas (GO), as researcher Moraes shows:

> the government report emphasised that the old capital was unable to grow physically due to its location between mountains in a narrow valley, the dry climate and the difficulties of water and sewage pipes, thus opposing the political ideals and Modern Brazil defended at the time by Getúlio Vargas' policy. (MORAES, 2003, p. 79 - 80).

---

[4] "It is the state of hygiene in which the urban and rural population lives, both in terms of its capacity to inhibit, prevent or impede the occurrence of endemic diseases or epidemics transmitted by the environment, and in terms of its potential to promote the improvement of mesological conditions favourable to the full enjoyment of health and well-being" (BRASIL, 2007, p. 14-15).

The author emphasises that the physical and economic characteristics of the Campinas region were in line with the ideas of the bourgeois revolution of 1930 and the political interests of Ludovico, who was looking for another regional area in which to implement his political ideas, contrary to the policies of the Caiados, who wanted the capital to remain in Vila Boa de Goiás (MORAES, 2003).

In Brazil, basic sanitation is a right guaranteed by the Federal Constitution of 1988 in art. 21 and defined by Law no. 11.445/2007 as the set of services, infrastructure and facilities for water supply, sanitation, urban cleaning, urban drainage, solid waste management and rainwater management (BRASIL, 2007).

Basic sanitation services are essential for promoting public health. Access to water in quantity and quality within the standards for human consumption and industrial use, respectively, is a factor in the prevention of various diseases, just as sanitation, public cleaning, solid waste management and urban drainage services are fundamental to the control and prevention of diseases associated with the inefficiency or absence of any of these services (LISBOA; HELLER; SILVEIRA, 2013).

Diseases related to the absence or inefficiency of basic sanitation can be classified into three broad categories: 1) faecal-oral transmission diseases; 2) diseases transmitted by insect vectors and 3) diseases transmitted by contact with water. Examples of diseases belonging to category 1 (one) are diarrhoea, enteric fevers and hepatitis A, while those belonging to category 2 (two) include dengue fever, yellow fever, leishmaniasis, malaria, Chagas disease and lymphatic filariasis. Category 3 diseases are schistosomiasis and leptospirosis.

Brazil follows a "developmentalist" model in which income and urban infrastructure are concentrated in large urban centres, industrial hubs or regions of economic interest. This development model intensifies regional socio-economic inequalities. For these reasons, significant social segments, especially the less favoured classes, end up suffering from precarious housing, health, environmental and economic conditions and a low quality of life (SILVEIRA, 2013). In Brazil there are 11.42 million people living in favelas, stilt houses or other irregular settlements, this figure corresponds to 6% of the country's population (Instituto Brasileiro de Geografia e Estatística - IBGE, 2010).

Against this backdrop, public intervention is needed to defend the environment, promote public health and improve sanitary conditions, with an emphasis on urban areas, where the majority of the Brazilian population is concentrated. Drawing up and implementing basic sanitation plans is fundamental to improving the living conditions of urban and rural dwellers. Before drawing up such a plan, it is necessary to know the real situation of urban infrastructure and household sanitation.

2.1 History and legal bases of sanitation

Concern about basic sanitation in Brazil began during the colonial period. With the arrival of the royal family, there was a significant population growth between 1808 and 1822, from 50,000 to approximately 100,000 people, according to Heller (1997). As a result, the demand for drinking water increased, as did the accumulation of waste and rubbish in the urban environment. During this period, sanitation actions were marked by the absence of the state, because the main interest was to guarantee the exploitation of the colony. Costa (2010) points out that the solutions were practically individual and the water supply was incipient, made individually, directly from surface springs or through some fountains that served a small portion of the population.

The 19th century saw the beginning of a phase characterised by the concession of water and sewage services from the state to the private sector. Although systems were built in several cities, the population served was restricted to those in the central centres, a system that is still widely used in Brazilian cities. The lack of quality in the services provided by the private companies sparked popular indignation, which led to large social demonstrations (BRASIL/MCIDADES, 2006). It was then that the state broke with the concessions to the private sector and once again took the lead in basic sanitation services, a theme that will be explored in greater depth in the city of Goiânia, especially in the third chapter. It was at this time that Saturnino de Brito, considered the father of National Sanitary Engineering, made a spectacular contribution to the construction of various sanitation systems in Brazil

(COSTA, 2003).

The 1950s and 1960s were marked by criticism of the financing mechanisms for water supply services, due to the lack of administrative autonomy (COSTA, 2003). The SAAEs (Autonomous Water and Sewage Services) and some state departments were then created in the form of local authorities. Some financing mechanisms were also created, through state funds and loans from state and federal banks (BRASIL/MCIDADES, 2006).

At the beginning of the 1970s, with urbanisation advancing, the National Basic Sanitation Plan (PLANASA) was launched, financed by the Severance Indemnity Fund (FGTS). The plan was a watershed for the sector and its goals included providing at least 80 per cent of Brazilian cities with a water supply and sewage system by 1980 (COSTA, 2003).

According to Madeira (2010), during the period in which the plan was in force, certain regions were favoured, causing profound regional inequalities in terms of basic sanitation services, which corroborated the failure of the sanitation model established by the 1971 plan.

The financial system created by PLANASA established two different models for organising sanitation services in the country: state and municipal service providers, as described by Heller (1996). A few years before PLANASA was launched, the state of Goiás had already created the Companhia de Saneamento de Goiás (SANEAGO) in 1967.

After the demise of PLANASA in 1985, there was some disarray in the sector due to the lack of a regulatory framework, which resulted in the following problems in Brazilian sanitation:

> Conflicts and competition between federal entities over ownership of services, the absence of a permanent federal investment policy, the refusal of municipalities to renew agreements with state operators signed during the PLANASA programme and the lack of definition of competences (MINISTÉRIO DAS CIDADES, 2004, p. 51).

During the period when PLANASA was in force and when it was abolished, the country went through a process of successive demonstrations and social struggles demanding improvements in access to health and sanitation services. As a result,

Law No. 8.080/1990 came into force, creating the Unified Health System (SUS) and making it an obligation to promote, protect and recover health through basic sanitation and health surveillance.

The concept of health established by this law includes factors such as food, housing, basic sanitation, education, income and work, leisure, among others. Basic sanitation has thus taken centre stage in public health policy (COSTA, 2010).

Sanitation issues have become relevant not only in the public health policy sector but also in urban policy. Article 2° of the City Statute, Law 10.257/2001, states that urban policy must guarantee the right to basic sanitation, and Article 3° includes basic sanitation in the list of duties of interest to urban policy:

> Art. 2°, I - guaranteeing the right to sustainable cities, understood as the right to urban land, housing and environmental sanitation. Art. 3, III - to promote, on its own initiative and in conjunction with the states, the Federal District and the municipalities, programmes for building housing and improving housing conditions and basic sanitation (...) IV - to establish guidelines for urban development, including housing, basic sanitation (...) (BRASIL, Law 10.257, 2001).

Basic sanitation in Brazil began to show new perspectives with the National Sanitation Policy, instituted by Law No. 11.445 of 2007, which filled a gap in the sector's specific legislation and defined guidelines, institutional arrangements and clear rules for planning, regulation, inspection, social control, universalisation of sanitation services and their management. This same law also established the universalisation of services for the supply of drinking water, sanitation, urban cleaning and solid waste management, drainage and urban rainwater management (BRASIL, 2007).

The definition of universalisation provided by the basic sanitation law indicates the likelihood that all people will achieve or have access to adequate sanitation services without any legal, economic, physical or cultural impediment, i.e. equal access for all citizens without any discrimination.

Article 17 of the law also stipulates that municipalities must draw up basic sanitation plans as a requirement for access to federal funding and resources (GALVÃO JÚNIOR et al, 2012). According to Lima Neto and Santos (2012), the new plans, unlike the previous ones, tend to be participatory and strategic planning instruments that will contribute to the more balanced development of municipalities. The preparation of these plans cannot be limited to technical-scientific knowledge alone, but must involve social, cultural, political, economic and environmental issues.

Article 52 of Law No. 11.445 of 2007 establishes that it is the Union's obligation to draw up the National Basic Sanitation Plan (Plansab) with a 20-year horizon and reviewed every four years. The plan must cover issues related to water supply, sewage disposal, solid waste and rainwater management and the provision of toilets and sanitation units for low-income populations (BRASIL, 2007).

2.2 Basic sanitation in Brazil

According to the World Health Organisation (WHO) and the United Nations Children's Fund - UNICEF (2015), only 68% of the world's population has access to adequate sanitation and more than two million people in the world live without adequate basic sanitation. In addition, more than 1.5 million children under the age of five die every year in the world due to problems related to inadequate water supply (WHO, 2015).

One of the great challenges of universalising sanitation is in relation to the rural population, since only 51% of this population has access to sanitation, while in the urban area coverage is 82%. In short, for every 10 people in the world, only three live in conditions of adequate sanitation (WHO and UNICEF, 2015).

In addition, more than 1 billion people around the world do not have access to a toilet and are forced to go to the toilet in the open air or on stilts, representing a serious public health problem. In the 1990s, 24 per cent of the world's population had no access to a toilet, while in 2015 this figure had fallen to 13 per cent.

In Brazil, 4 million people do not have a toilet (Progress *on Sanitation and Drinking-Water,* 2015 - WHO/UNICEF).

Figure 2 refers to the population's access to water and sanitation.

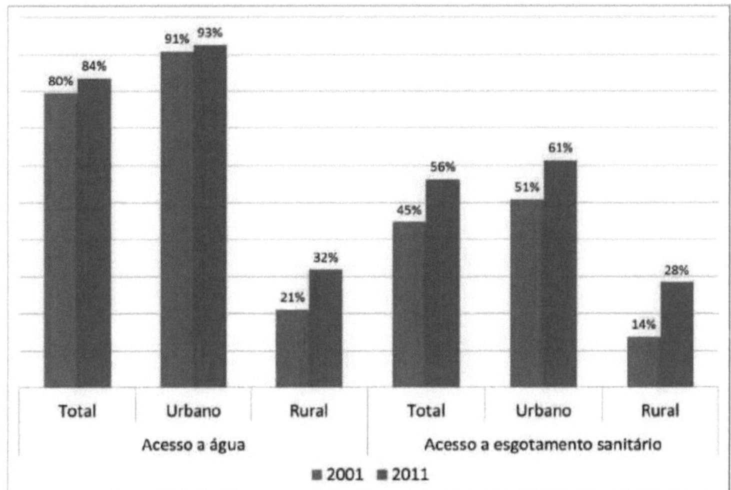

Figure 2 - Percentage of the Brazilian population with access to water and sanitation*
Source: PNAD/IBGE (2001 and 2011).
*Access to water: percentage of residents in households with a general water supply network. Access to sewage: percentage of residents in urban households with a sewage collection system and percentage of rural residents in households with a sewage collection system or septic tank.

Figure 2 shows that only 56 per cent of the total Brazilian population (urban and rural) has access to a sewage system and 84 per cent to treated water. Data from the National Basic Sanitation Survey (PNSB, 2009) show that around 30 per cent of the sewage collected in Brazil is treated.

Among the deficiencies in sanitation there is the problem of loss of treated water: for every 100 litres of water collected and treated, on average only 67 litres are consumed, i.e. 37% of treated water in Brazil is lost through leaks, clandestine connections, lack of measurement or incorrect measurement of water consumption. The North region has the highest rate of water loss in the country (47 per cent) and the Southeast has the lowest rate, approximately 32 per cent (PNAD, 2011).

Figure 3 shows that more than 30 million Brazilians don't have access to a quality drinking water supply, 85 million don't have proper access to sewage, a further 130 million don't benefit from sewage treatment and, surprisingly, there are still 6.6 million Brazilians who don't even have a toilet.

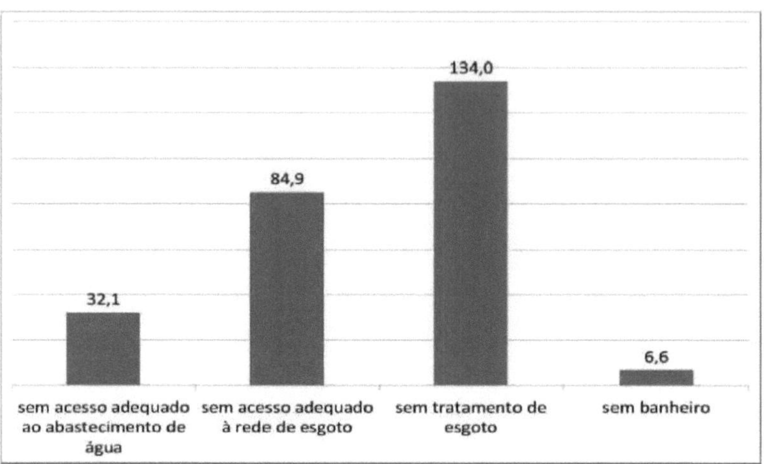

Figure 3 - Number of Brazilians without adequate access to water supply, sewerage and toilets (millions of people)* Source: PNAD/IBGE (2011).
*Adequate access to water: percentage of residents in households with a general water supply network. Adequate access to sanitation: percentage of residents in urban households with a sewage collection system and percentage of rural residents in households with a collection system or septic tank.

The evolution of the service exists, but it is still slow.
Data from the 2011 National Household Sample Survey (PNAD) shows that between 2001 and 2011, domestic sewage collection increased by just 11 percentage points (p.p). This rate is lower than the growth of other public utilities in the same period, such as telephony, which rose from 58 per cent in 2001 to 91 per cent in 2011. In a similar period, electricity, which reached 95.5% of households in 1991, evolved to a situation of near universalisation (99.3%) (PNAD, 2011).

The rate of basic sanitation coverage in Brazil is still very uneven between regions. Data from the National System of Sanitation Indicators (SNIS, 2011) shows that the only federal units with a total water service index - which includes urban and rural areas - above 90 per cent are São Paulo, the Federal District and Mato Grosso do Sul (see Figure 4).

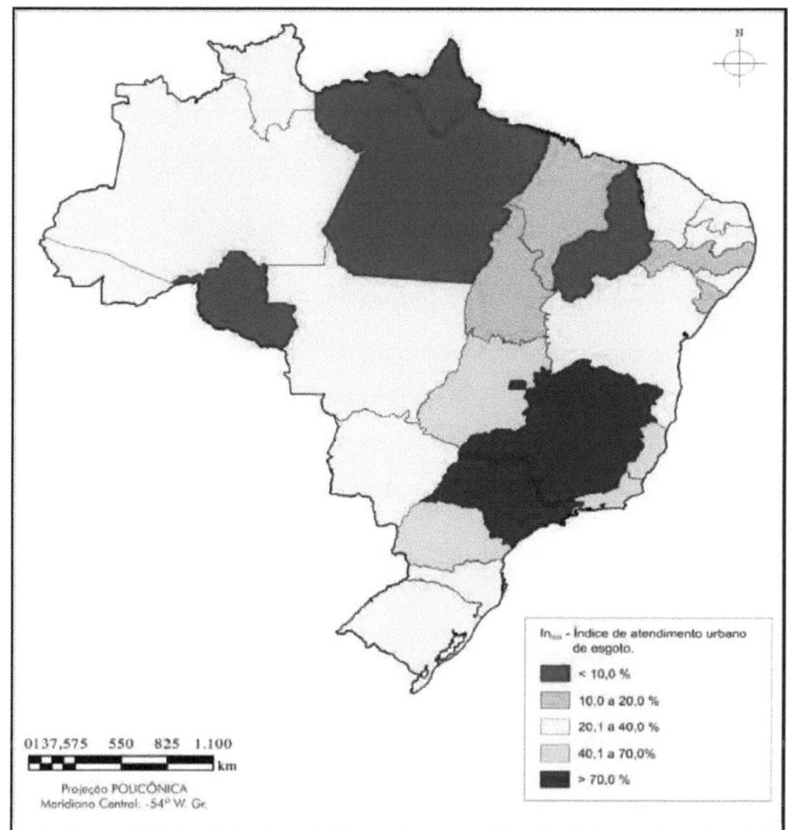

Figure 4 - Heterogeneity of urban sewage service in Brazilian states. Source: SNIS, 2011.

Regarding total sewage service, the only state with a rate above 70% is São Paulo. The states of Amapá, Pará, Rondônia, Acre and Alagoas had the lowest coverage rates, with no more than 60 per cent. With regard to urban sewage service, only São Paulo, Minas Gerais and the Federal District had rates above 70 per cent. The states with the worst rates (less than 10%) were Rondônia, Pará, Amapá and Piauí. The basic sanitation situation in Brazil's rural areas is even more critical and the state has been unable to achieve universal sanitation through existing public instruments and policies. According to PNAD 2011, around 70 per cent of rural inhabitants do not have access to adequate sanitation.

Table 1 shows the evolution of urban households with access to water supply, sewerage or septic tanks between 2000 and 2010.

Table 1 - Proportion of urban households with a general water supply, sewage system and septic tank, by major region - 2000/2010.

REGIONS	Water Supply (%)		Sewerage and septic tank (%)		Rubbish collection (%)	
	2000	2010	2000	2010	2000	2010
North	62,5	66,2	46,7	40,6	77,6	93,6
North-East	85,5	90,5	51	56,7	82,4	93,7
South East	94,6	95,3	87,8	90,7	96,4	98,8
South	93,4	94,7	72,6	78,2	97,2	99,3
Centre - West	82,4	90	45,9	56,3	92,5	98,4
Brazil	**89,8**	**91,9**	**72**	**75,3**	**92,1**	**97,4**

Source: Prepared by the author based on data from IBGE, Demographic Census 2000/2010.

Table 1 shows that basic sanitation infrastructure in urban households has improved over the period evaluated. The underdeveloped regions have made progress in the provision of sanitation services over the last 10 years, especially in terms of rubbish collection and water supply, but this has not yet been enough to reduce regional inequalities.

There has been growth in the water supply service by general network in all regions of Brazil, albeit to a lesser extent.

unequal. In 2010, the Southeast and South continued to have the highest percentages of households with a general water supply network, 95.3 per cent and 94.7 per cent respectively. On the other hand, the North region, despite progress, remained with the lowest percentage in the country (66.2 per cent). The Northeast and Centre-West regions advanced to 90% coverage, which was the exclusive privilege of the South and Southeast in 2000. It can be said that at a national level, there has been no considerable progress, in the 10 years evaluated, the evolution was only 2.1 percentage points (p.p.) in this indicator.

With regard to rubbish collection, all Brazilian regions also showed growth between 2000 and 2010, with the North of the country standing out for having shown growth of 16 p.p. However, having rubbish collection does not mean that municipalities are providing environmentally appropriate waste disposal and adequate disposal of waste. Brazil still faces major challenges in terms of solid waste, three of which stand out in brief: eradicating rubbish dumps and controlled landfills in municipalities, implementing and enforcing municipal selective waste collection and enforcing the reverse logistics of waste defined by the National Solid Waste Policy

(PNRS). These objectives can be achieved by starting to draw up Municipal Plans for Basic Sanitation and Solid Waste.

Of the three sanitation conditions shown in Table 1, sewage is the indicator with the worst percentages, which means that Brazilian municipalities have many challenges to overcome before they can achieve satisfactory rates that improve the population's housing and health conditions. There was an increase in the proportion of households connected to the general sewage network or with a septic tank in four of the country's five regions. The North saw a drop of 6.1 percentage points, while the Northeast saw an increase of 5.7 percentage points. The Southeast remained the region with the best sanitation conditions in urban households, followed by the South. The Centre-West surprised us with the biggest increase in the period, rising by more than 10 percentage points.

Although a significant part of the Brazilian population, according to PLANSAB (2013), has adequate access to drinking water supply and solid waste management, the deficiencies in sanitation services are still significant in general. The state, through its public policies , has not managed to universalise basic sanitation services for the population.

An overview of the basic sanitation situation in Brazil will be presented below, based on PLANSAB (2013). By analysing a few selected variables, it will be possible to get an overview of the reality of basic sanitation and the deep socio-economic and regional inequalities that exist in the country.

2.2.1 Water supply

PLANSAB (2013) establishes three forms of water supply, namely: 1) service through a general network with internal channelling 2) a well or spring and 3) a general network channelled through the property itself. Approximately 85 per cent of the Brazilian population are served by the first option and 11 per cent by wells or springs

In proportional terms, the inhabitants of the North are the ones who most use water from wells or springs to meet their daily needs. The population of the Northeast and Southeast regions has the highest proportion of households connected to the mains, while in the Centre-West region there is still a significant proportion of households served by cisterns, water tankers and/or other forms of water supply (6.6%) (PLANSAB, 2013).

The biggest water supply deficits are to be found in the north-east and north, where 94 and 100 per cent of their inhabitants, respectively, meet their water needs inadequately.

Another important point to highlight is the socio-economic conditions of the people living in unsuitable water supply conditions. Almost 70 per cent of the population included in the deficit of access to water supply have a monthly income of up to % of the minimum wage per resident. Data from PLANSAB (2013) also shows that the lower the education level of the population, the more vulnerable they are to inadequate water supply.

Another important aspect is the quality of the services provided by . In 2010, 38 million Brazilians were supplied with water from 1,046 public water supply systems, which did not fully meet the potability standards established by Ordinance No.
2.914/2011 of the Ministry of Health (MS) (SISAGUA 2010, 2011 and MINISTÉRIO DA SAÚDE, 2013). It is not enough just to achieve universalisation of this service, but above all attention must be paid to the quality of the water distributed, which must meet the standards of potability established by current legislation.

2.2.2 Sanitary sewerage

The data presented by PLANSAB (2013) shows that 53 per cent of the total volume of sewage generated in Brazil is disposed of via the general sewage network or rainwater network, 12 per cent via septic tanks, 26 per cent via rudimentary tanks, 6 per cent to ditches, rivers, lakes, the sea or other destinations and the remaining 3 per cent is due to the lack of a toilet.

The deficiencies in sanitary sewerage services are reflected in the proportion of the population that does not have access to sewage disposal and/or treatment, in addition to households that do not have toilets and that discharge effluent directly into drains.

Figure 5 shows the distribution of the different waste disposal practices adopted in each region. It shows that the regions with the greatest proportional deficit are: Centre-West (50%), Northeast (56%) and North (69%). Among the inadequate waste disposal practices that have the greatest negative impact on the environment and the health of the population are rudimentary cesspools and the direct disposal of waste into watercourses and soil.

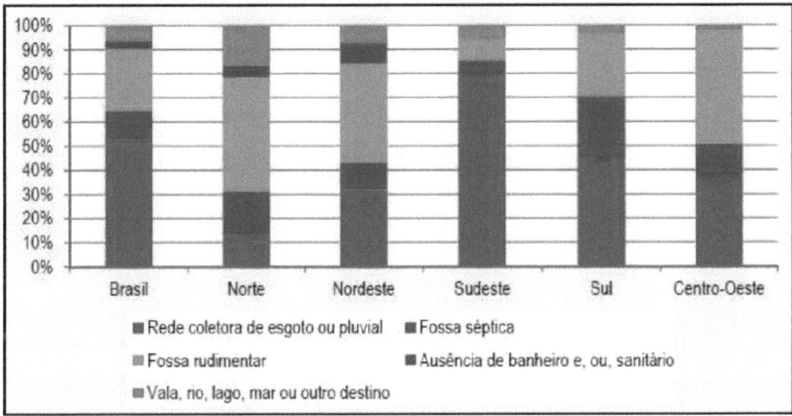

Figure 5 - Practices adopted for sewage disposal, as a proportion of the population by macro-region and Brazil, 2010.
Source: PLANSAB (2013, p. 30).

Regional inequalities in terms of sewage disposal in Brazil are more visible when analysing information on sewage collection and treatment separately. Table 2 shows the percentage of municipalities by region that had sewage collection and treatment in 2008.

Table 2 - Total and percentage of municipalities with sewage collection and treatment, by major region - 2008

MAJOR REGIONS	MUNICIPALITIES		
	Total	With sewage collection (%)	With sewage treatment (%)
North	449	13	8
North-East	1793	46	19
South East	1668	95	48
South	1188	40	24
Centre-West	466	28	25
Brazil	5564	55	29

Source: Prepared by the author based on data from IBGE, National Basic Sanitation Survey, 2008.

While the northern region has only 13 per cent of municipalities with sewage collection and 8 per cent with sewage treatment.

The Southeast has 95 per cent of municipalities with a sewage collection system and 48 per cent with treatment. At national level, just over half of Brazil's municipalities (55%) have a sewage collection system and only 29% have treatment. The North region has the worst sanitation scenario, with only 14.36 per cent of the total sewage collected being treated, followed by the Northeast region with 28.8 per cent of sewage treated. The regions Southeast, South and Centre-West have the best sewage treatment rates, but none of them reach 50% treatment (SNIS 2014). Brazil has a long way to go to reach 100 per cent of its municipalities with sewage collection and treatment systems.

Expanding the discussion a little further, data from PLANSAB (2013) presents some relevant information on sanitation. More than 48 per cent of the Brazilian population does not have access to sewage collection. There are more than 3.5 million people spread across the country's 100 largest cities, dumping sewage irregularly even though collection networks are available.

In economic terms, the cost of universalising access to the four basic sanitation services (water, sewage, solid waste and urban drainage) in Brazil between 2014 and 2033 would be around R$500 billion. To universalise drinking water and sewage alone, this cost will be R$303 billion over twenty years (PLANSAB, 2013).

2.2.3 Solid Waste Management

Once again, Brazil's challenge lies in rural areas: while 90% of urban households had rubbish collection in 2010, 72% of rural households had no waste collection at all (IBGE, 2010). The urban population with adequate access to the domestic solid waste (DSW) collection service exceeds 80 per cent, while in the rural population, coverage does not reach 30 per cent. Data from the PNAD (2011) show that a further 38 million people do not have a collection service for household solid waste. Among the population served by this service, the highest coverage is once again in the Southeast (89.5 per cent), while the Northeast has the lowest coverage (63.2 per cent).The practice of burning or burying waste is still common among the portion of the population with a shortage of solid waste collection, as shown in Table 3.

Table 3 - How solid waste is disposed of by the population in a situation where there is a shortfall in the disposal of solid waste - 2010.

Brazil Macro-regions	Proportion of the population (%)			
	Burn or bury	Indirect collection (urban environment)	Waste ground and public places	Water bodies and others
North	69,8	21,0	6,6	2,6
North-East	55,5	26,4	16,7	1,5
South East	59,9	32,7	3,5	3,9
South	67,1	25,4	1,8	5,7
Centre-West	43,4	48,9	5,1	2,7

Source: Prepared by the author based on data from the IBGE - Demographic Census (2010).

The Northeast region has the largest population with a waste collection deficit (19,456,791 thousand people). Of this total, almost 70 per cent choose to bury or burn rubbish on their properties, as is the case in other regions (PLANSAB, 2013).

With regard to the final disposal of waste, data from the PNSB (2008) show that dumps are still part of the reality in 48 per cent of Brazilian municipalities. A survey by SNIS (2010) shows that of the 1,429 waste disposal units on the ground, only 37 per cent were declared by managers as sanitary landfills, 28 per cent as controlled landfills and 35 per cent as dumps. It is also possible that parts of these categorisations were not made in the appropriate categories, since the information is self-declared by municipal managers.

Another relevant aspect refers to the presence of waste pickers working in rubbish dumps. More than 13 per cent of the municipalities that took part in the PLANSAB survey (2013) - 286 municipalities out of a total of 2,070 - reported the presence of waste pickers at waste disposal sites. This is exacerbated when the waste pickers are children and adolescents. In addition, these people are exposed to various risks of accidents and unhealthy conditions that can lead to their death or the contraction of various serious illnesses.

2.2.4 Urban drainage

Due to the complexity of this issue, the technical team that prepared PLANSAB (2013) decided to address the drainage and urban rainwater management component based on the records of flooding problems in the 2008 PNSB. The data showed that 2,257 municipalities were affected by one or more flood events, and of this total, 1,862 declared that the rainwater management system had been expanded or improved in 2008 (IBGE, 2009).

The Southeast had the highest number of records of flooding, accounting for 52 per cent of all municipalities, followed by the South and Northeast with 43 per cent and 40 per cent of municipalities, respectively, reporting problems. The North and Centre-West regions had the lowest proportion of municipalities with floods or inundations, Figure 6.

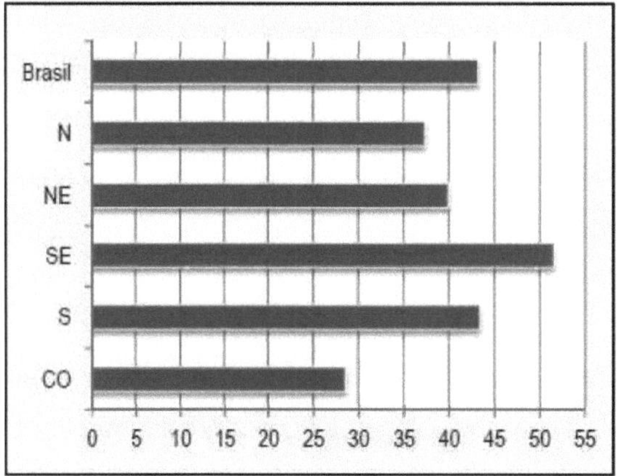

Figure 6 - Proportion of municipalities with flooding over a five-year period. Source: PNSB (IBGE, 2008 p. 40).

Table 4 shows that 82 per cent of municipalities with between 100,001 and 500,000 inhabitants took part in the survey in 2008 and 93 per cent of them reported problems with flooding. Similarly, all the municipalities with more than 500,000 inhabitants that took part in the survey also reported problems with flooding.

Table 4 - Occurrence of floods in the five-year period by population group

POPULATION RANGE (inhab.)	MUNICIPALITIES				
	Total	Participants in PNSB 2008		With flooding	
	(N°)	(N°)	(%)	(N°)	(%)
Up to 5,000	1.257	1.257	100	279	22
From 5,001 to 20,000	2.664	2.505	94	998	40
From 20,001 to 100,000	1.371	1.196	87	770	64
From 100,001 to 500,000	233	192	82	178	93
More than 500,000	40	32	80	32	100

Brazil	5.565	5.182	93	2.257	44

Source: Adapted by the author based on PNSB (IBGE, 2008).

One of the factors behind the occurrence of floods in large cities is the high level of soil sealing combined with the inefficiency of the rainwater drainage and management system.

Even though the soil sealing index is lower in cities with smaller populations, Table 4 shows that these municipalities also had significant proportions of cities with flooding. This can be explained by the inefficiency or non-existence of drainage and rainwater management combined with the relief conditions of each city. At a national level, 44% of the Brazilian municipalities that took part in the survey still suffer from flooding, ranging from loss of property to deaths and socio-economic problems.

The number of floods in Brazilian cities has been increasing in recent years. Among the many damages, these floods can minimise the population's quality of life and cause damage to public and private property, destroying crops, degrading the environment, altering the natural and artificial landscape, causing human death and increasing the risk of transmitting diseases associated with rainwater.

The scarcity of risk management tactics and emergency and contingency actions has exacerbated these problems. In addition, the absence and/or inefficiency of urban and environmental planning, uncontrolled urbanisation, large-scale sealing of urban soil, irregular occupation of environmentally protected areas, such as valley bottoms, are some of the factors that trigger urban flooding.

CHAPTER 3

BASIC SANITATION STRUCTURE IN GOIÂNIA AND HEALTH

3.1 Overview of sanitation services provided by SANEAGO

a) History

The history of sanitation in Goiás, especially in Goiânia, began with the company Saneamento de Goiás-SANEAGO. The company was founded in 1967 under State Law 6680/67 with the aim of distributing treated water and collecting and treating domestic effluent.

During the 1960s and 1970s, the sanitation sector in Brazil did not receive enough attention from the government. During the 1960s, there was the first attempt by the federal government to solve the problem, the "Three-Year Programme Method". The National Housing Bank already existed and, with it, three superintendencies were created: Superintendência Financeira da Habitação (SFH), Superintendência Financeira do Saneamento (SFS) and Superintendência Financeira de Desenvolvimento Urbano (SFDU), as reported in the previous chapters. At this time, the federal government began approving the feasibility studies of the State Sanitation Companies and the projects resulting from the implementation of the National Sanitation Plan (PLANASA).

In Goiás, Governor Pedro Ludovico Teixeira awarded the concession to operate the water and sewage systems to Cia. Melhoramentos S.A. for 30 years. This concession was reviewed and cancelled under Jerônimo Coimbra Bueno. The sanitation sector was then managed by the Department of Roads and Public Works until the State Sanitation Department (DES) was set up under Mauro Borges.

As a requirement of the National Housing Bank's Sanitation Financial System (BNH/SFS), Saneamento de Goiás S.A. (SANEAGO) was created on 13 September 1967, during the government of Otávio Laje. It was set up on 29 June 1969. Since then, Saneago has been responsible for studies and projects, the construction of water supply and sewage systems, as well as the operation and maintenance of the systems set up in Goiás. Today, the company operates in 225 municipalities and has service rates of 96 per cent for water and 51.9 per cent for sewage.

b) SANEAGO: water and sewage

Law No. 11.445 of 2007 made it compulsory to draw up municipal basic sanitation plans as a condition for accessing federal budget funds earmarked for the sector. The deadline for submitting the plans was extended to 31 December 2017. This plan has not yet been drawn up for the municipality of Goiânia, but a call for tenders has already been issued for Public Tender No. 003/2015 to hire specialised technical services to draw up the document (GOIÂNIA, 2015).

The water sources that supply the population of Goiânia are the Meia Ponte River and the João Leite Stream. There are three water treatment plants in Goiânia: ETA Eng. Rodolfo José da Costa (ETA Meia Ponte) and ETA Jaime Câmara and more recently ETA Mauro Borges (Figure 7). The Meia Ponte WTP collects water from the Meia Ponte River and supplies around 50% of Goiânia's population. The Jaime Câmara treatment plant was inaugurated in 1957 with a capacity to treat around 150 l/s, which meant treatment for 49% of the municipality's population.

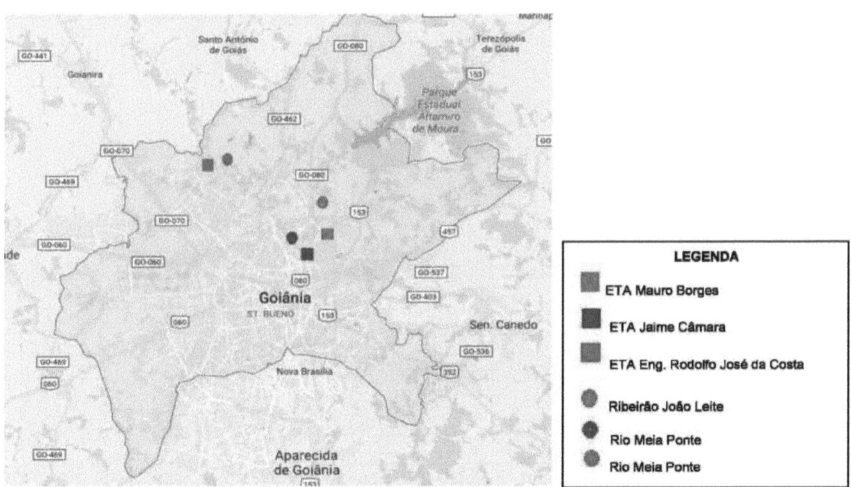

Figure 7 - Map of Goiânia showing the location of the WTPs and water sources.
Source: Google Earth on 10/01/2017, modified by the author.

In 1988, the João Leite System began operating, serving 51 per cent of the population of Goiânia and the metropolitan region. As the population grew, the amount of treated water was not enough to meet demand, which led to the construction of the Mauro Borges Producer System, which consists of the Dr Henrique Santillo Dam, a Raw Water Pumping Station and the Mauro Borges Water Treatment Plant (Figures 8 and 9).

Figure 8 - Dr Henrique Santillo Dam (a) and Raw Water Pumping Station (b) of the Mauro Borges Producer System, Goiânia- GO.
Source: SANEAGO website, 2016.

The Mauro Borges System is designed to produce 6,000 litres of water per second, enough to meet the demand of part of the population of Goiânia and the entire metropolitan region until 2040 (SANEAGO, 2016).

Figure 9 - Mauro Borges Water Treatment Plant, Goiânia- GO. Source: SANEAGO website, 2016.

Goiânia has a sewage treatment plant called ETE Dr. Hélio Seixo de Britto

(Figure 10), which was inaugurated in 2003 with the capacity to treat up to 2,300 litres of sewage per second. The average flow of sewage currently arriving at the plant is around 1,400 litres per second.

Figure 10 - Dr Hélio Seixo de Britto Water Treatment Plant. Source: SANEAGO website, 2016.

This sewage treatment plant carries out advanced primary treatment (APT) or chemically assisted primary treatment (CEPT), which is based on the removal of suspended solids through the physical-chemical processes of coagulation, flocculation and sedimentation. The sewage goes through the removal of coarse materials, sand and after the addition of coagulants and polymers, it is possible to reduce 50 per cent of the organic load and 80 per cent of the total suspended solids. The Goiânia sewage treatment plant treats approximately 75 per cent of the sewage collected in the municipality and its contributing basins are the Anicuns stream and its tributaries (Macambira, Cascavel, Vaca Brava, Capim Puba and Botafogo), the Caveirinha and Fundo streams and the João Leite stream (SANEAGO, 2010).

Despite having a significant percentage of its sewage collected and treated, the Goiânia WWTP still discharges sewage with a high organic load and pathogenic organisms into the receiving body, because the effluent has not yet received secondary or tertiary treatment. Secondary treatment is aimed at the biological degradation of carbonaceous compounds and the reduction of nutrients such as nitrogen and phosphorus, which in excess can cause the phenomenon called eutrophication. In tertiary treatment, in addition to removing nutrients that were not sufficiently removed in secondary treatment, the effluent undergoes a disinfection process aimed at removing pathogenic organisms. The federal government, through the Ministry of Cities, has made financial resources available for the expansion and implementation of secondary treatment at the Goiânia WWTP, but the work has not yet been finalised.

3.2 Analysing basic sanitation services in Goiânia

Table 5 gives a general overview of the conditions of permanent private households and the residents of these households, with regard to the existence or not of culverts or manholes, open sewers and rubbish accumulated in the streets.

Table 5 - Conditions of permanent private households and residents of Goiânia-GO - 2010

Features	No. of households	%	No. of residents	%
Manhole				
There is	222.637	53,0	670.962	51,96
It doesn't exist	196.803	46,8	617.800	47,85
No declaration	787	0,2	2.437	0,19
Open sewage				
There is	2.054	0,5	6.663	0,52
It doesn't exist	417.386	99,3	1.282.099	99,30
No declaration	787	0,2	2.437	0,19
Rubbish accumulated in the streets				
There is	10.801	2,6	35.008	2,71
It doesn't exist	408.639	97,2	1.253.754	97,10
No declaration	787	0,19	2.437	0,58
Total	420.227		1.291.199	

Source: Prepared by the author using data from the 2010 IBGE Demographic Census.

With regard to rainwater drainage, almost 50 per cent of households and residents are not adequately served by this service. Rainwater drainage is an extremely important sanitation service, the lack of which can lead to flooding, which in turn is responsible for the proliferation of insect vectors and diseases such as leptospirosis. Figures 11, 12 and 13 show some points of flooding in the municipality of Goiânia.

Figure 11 - Points of flooding during the rainy season in Vila Roriz, a suburb of Goiânia - GO. Source: Diário de Goiás website, 2016.

The problem of flooding is not just in the outlying neighbourhoods of Goiânia. In the rainy season of 2015 and early 2016, heavy rains caused flooding in various parts of the capital (Figure 13).

Figure 12 - Flooding on the small viaduct between Flamboyant and Jardim Goiás, under Avenida Jamel Cecílio (a) and flooding on the Avenida 85 Viaduct (b) in Goiânia - GO. Source: Diário de Goiás website, 2016.

The central region of Goiânia is the place most prone to flooding according to research carried out by Luiz (2012), because in this region, the drainage structure does not efficiently capture the volume of surface runoff, in addition, the central region is close to the streams which, due to the relief and slope tend to accelerate the flow of water towards the drainage channels, which contributes to flooding in some parts of the city, as can be exemplified in the vicinity of the Marginal Botafogo (Figure 14).

Figure 13 - Flooding of the Marginal Bota Fogo (a) and the Jaime Câmara Tunnel near Mutirama Park (b) in Goiânia- GO.
Source: Globo's G1 website, 2016.

The areas close to the Meia Ponte River and Anicuns Stream plain, where some of the city's streams flow, are also places where flooding is evident, due to the reduction in water infiltration capacity in the soil, affected by sealing and drainage channelling (LUIZ, 2012).

Table 6 shows the sanitation conditions of Goiânia residents from the 1991 Demographic Census to 2010. For reference purposes, the population of Goiânia, according to the 1991, 2000 and 2010 Censuses, was 922,222 thousand people, 1,093,007 and 1,302,001 million, respectively.

Table 6 - Residents in permanent private households, by type of sewerage system in Goiânia - GO according to the 1991, 2000 and 2010 Censuses.

Type of sanitary facilities	1991		2000		2010	
	No. of residents	(%)	No. of residents	(%)	No. of residents	(%)
General sewage or rainwater system	669.121	73,00	797.492	73,33	873.082	67,36
Septic tank	24.954	2,72	60.929	5,60	112.129	8,65
Rudimentary cesspit	198.949	21,71	211.182	19,42	305.973	23,61
Ditch, river or lake	1.771	0,19	9.087	0,84	3.460	0,27
Another outlet	7.942	0,87	2.183	0,20	886	0,07
No sanitary facilities	13.816	1,51	6.706	0,62	660	0,05
Total	916.553	100	1.087.579	100	1.296.190	100

Source: Organised by the author based on the IBGE Demographic Censuses of 1991, 2000 and 2010.

There has been a significant increase in population over the last 19 years, but the sanitation infrastructure has not been able to keep up with this growth. It can be seen that in 1991, 73 per cent of residents were served by a general sewage or rainwater system. Almost 10 years later, the population increased and the service rate remained stagnant. In 2010, coverage fell to 67.3 per cent of residents with a general sewage system and the rate of residents with a cesspit rose from 21.7 per cent in 1991 to 23.6 per cent in 2010.

Adding up the number of residents with inadequate sanitation facilities, which includes people who dispose of their effluent in rudimentary pits, ditches, rivers or lakes, other types of drains and those without any type of facility, in 1991 and 2010, gives a total of 222,478 and 310,979 thousand residents, respectively. In percentage terms, there was a practically insignificant reduction from 24.2 per cent in 1991 to 23.9 per cent in 2010,

According to the 2010 Demographic Census, there are 420,227 households in Goiânia. Figure 14 shows the sanitation conditions of these households.

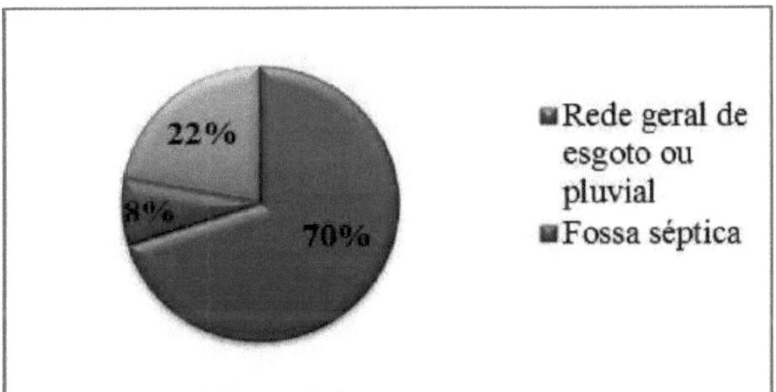

Figure 14 - Permanent private households by type of sewerage system, Goiânia - GO in 2010.
Source: Prepared by the author using data from the 2010 IBGE Demographic Census.

Of all the households, 22% are still not served by a general sewage system or rainwater, i.e. they still dispose of their domestic effluent in rudimentary pits, ditches, rivers or lakes. Only 279 households did not have any kind of sanitary installation. The collection, treatment and disinfection of domestic sewage is essential for protecting public health and preserving the environment. Sewage contains numerous living organisms such as bacteria, viruses, worms and protozoa, most of which are released along with human waste. Many infections can be transmitted from a sick person to a healthy one via human excreta. Domestic sewage can contaminate water, food, household utensils, soil, water sources and can be carried by vectors such as flies and cockroaches, which causes new infections (BRAGA et al., 2005). Data on water supply can be found in Figure 15.

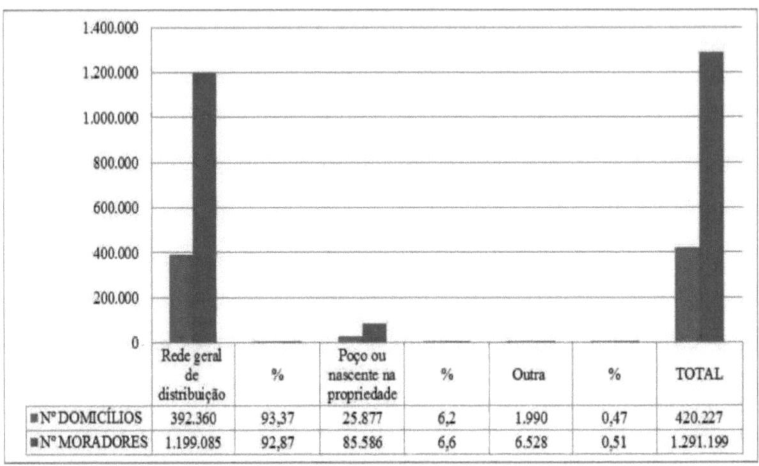

Figure 15 - Water supply in Goiânia - 2010.
Source: IBGE - 2010 Census. Prepared by the author.

Regarding water supply, more than 93 per cent of households and 92 per cent of residents in Goiânia were served by a general water distribution network and around 6 per cent were supplied by wells or springs on the property, according to the latest IBGE Demographic Census of 2010. More recent information on water supply and sanitation in Goiânia is shown in Table 7.

Table 7 - Main basic sanitation indicators in Goiânia - 2014.

Water service indicator (%)	Total	99,62
	Urban	100
Sewage service indicator (collection) (%)	Total	84,3
	Urban	84,62
Indicator of sewage treated per water consumed (%)	64,72	
Investment 5 years (Millions R$/year)	671,93	
Revenue 5 years (Billions R$/year)	2.638,24	
Investment/revenue (%)	25,47	
New water connections	33.313	
Missing connections for universalisation	35.284	
New sewage connections	18.264	
Missing connections for universalisation	75.946	
Distribution losses indicator 2014 (%)	21,07	
Average rate (R$/m³)	3,58	

Source: National Sanitation Information System - SNIS (2014). Prepared by the author.

The total water/sewage service indicator refers to the urban and rural population served by water supply and/or sewage collection. In the case of Goiânia, 99.62% and 84.3% of the total population (urban and rural) are served by water supply and sewage collection, respectively. With regard to water loss in distribution, Goiânia is the fourth municipality with the lowest loss rate (21.07%), behind only Limeira-SP (14.08%), Ribeirão Preto-SP (15.89%) and Santos-SP (18.98%) (SNIS, 2014).

The average for Brazilian municipalities in terms of water and sewage collection is 83% and 49.8%, respectively, which means that although Goiânia has not yet achieved universalisation of these services, the capital of Goiás stands out for having an index above the total average for Brazil (SNIS, 2014). It is also worth noting that only 23 municipalities in total have universalised water services and only two have 100% sewage collection (Franca - SP and Belo Horizonte - MG) and (TRATA BRASIL, 2016).

Only 64 per cent of the sewage generated by the population of Goiânia (based on the volume of water consumed) receives treatment. This is above the average for all Brazilian municipalities (40.8 per cent), according to SNIS (2014), but it can still be considered low. Among the indicators presented so far, sewage treatment is the main challenge for Goiânia in the quest for universalisation of basic sanitation services, and can be considered the main bottleneck to be overcome. There are still 75,946 missing sewage connections and 35,284 missing water connections to universalise water and sewage services in the municipality of Goiânia.

Another relevant piece of information that deserves to be highlighted about sanitation in Goiânia is the relationship between investment and revenue. Only 25 per cent of the total amount collected in the last five years in Goiânia was invested (BRAGA et al., 2005). Of all the municipalities included in the survey, more than 60 per cent invest less than 30 per cent of the amount collected. The municipality that won first place in the overall ranking, Franca-SP, invested 41.61% of the amount collected; other municipalities such as Praia Grande-SP, Caxias do Sul-RS and Vitória-ES invested 90.7%, 60.18% and 81.69%, respectively (SNIS, 2014). It should be noted that all operating revenue is considered to be collected, both the investment made by the provider and those made by the public authorities (TRATA BRASIL, 2016).With regard to sanitation indicators, Table 8 shows the evolution of coverage indicators in Goiânia between 2010 and 2014.

Table 8 - Evolution of sanitation coverage indicators in Goiânia between 2010 and 2014.

CAPITAL	INDICATOR	YEARS					EVOLUTION (P-P)
		2010	2011	2012	2013	2014	
Goiânia - GO	Total water service (%)	99,61	99,62	99,62	99,62	99,62	0,01
	Total sewerage service (collection) (%)	76,64	76,42	79,8	79,48	84,3	7,66
	Sewage treatment (%)	64,32	62,73	61,34	63,45	64,72	0,4
	Distribution losses (%)	23,47	23,54	22,17	21,31	21,07	2,4
	Evolution in investments R$ (MM)	118,6	113,5	135,2	132,9	171,5	-

Source: SNIS (2014). Prepared by the author.

Goiânia is one of the 10 capitals with the worst performance in terms of sewage treatment between 2010 and 2014 and one of the top 10 with the best performance in terms of reducing water loss in distribution (TRATA BRASIL, 2016; SNIS, 2014).

The average annual investment in sanitation in Goiânia was very significant, at R$134.39 million, totalling an average annual investment per inhabitant of R$19.03. Among the capitals, Vitória-ES had the highest average annual investment per inhabitant (R$65.19), but did not achieve good progress in the total sewage service indicator, growing only 3.49 percentage points, and negative progress in total water service (SNIS, 2014).

With regard to solid waste, Goiânia has voluntary drop-off points (PEV) for recyclable materials and the Goiânia Selective Collection Programme (PGCS), which was set up in the municipality by Municipal Decree No. 754 on 28 March 2008 (Figure 16).

Figure 16 - Trucks used to collect recyclable materials (a) and Voluntary Delivery Point (PEV) (b) in Goiânia-GO.
Source: Goiânia City Council website.

Table 9 shows how rubbish is disposed of by number of residents, according to data from the 1991, 2000 and 2010 Censuses.

Table 9 - How the population of Goiânia - GO disposes of its rubbish.

How rubbish is disposed of	Year		
	1991	2000	2010
1 Collected	850.906	1.073.346	1.293.603
1.1 Collected by cleaning services	841.158	1.031.303	1.216.982
1.2 Cleaning service bucket collector	9.748	42.043	76.621
2 Burned (on the property)	31.779	8.572	1.919
3 Buried (on the property)	4.736	1.948	219
4 Dumped in waste ground or pond	27.211	3206	115
5 Other destination	1.921	507	334
Total	916.553	1.087.579	1.296.190

Source: Organised by the author based on the IBGE Demographic Censuses of 1991, 2000 and 2010.

In 1991, of the total number of residents (916,553,000), more than 850,000 had access to rubbish collection (92.8%), of which 98.8% had their waste collected by a cleaning service and only 1.2% by a cleaning service skip. The rest of the residents (65,647,000) disposed of their waste in other ways, namely by burning, burying, dumping and/or other means. The situation improved in 2010, with 99.8 per cent of residents covered by cleaning services. There has been a reduction both in absolute figures and in the percentage of residents who bury, burn, dump or otherwise dispose of their waste. However, when you leave the numbers aside and look at the reality, especially in the outlying neighbourhoods, it is still very common for residents to throw rubbish on vacant lots (Figure 17).

Figure 17 - Urban solid waste dumped in vacant lots in the Portelinha neighbourhood in Goiânia - GO
Source: TV GUAIAMUM website, 2016.

The company responsible for collecting and disposing of solid urban waste in Goiânia is COMURG. According to information from the company, the monthly average of solid waste (domestic and public) collected in the municipality is 35,000 tonnes and over 70,000 tonnes of construction and urban cleaning waste. Table 10 gives an overview of the amount of solid waste collected in the municipality.

Table 10 - Solid waste collected by COMURG in Goiânia from 2008 - 2012

YEAR	Total solid waste* (tonnes)	Total Health Service Waste (tonnes)	Total recyclable materials collected (tonnes)	Number of tyres removed (unit)	Total solid waste* received at the landfill (tonnes)
2008	409.530,41	2.477,14	596,26		409.530,41
2009	420.817,82	2.574,55	3.284,78	172.750	420.817,82
2010	421.483,83	2.636,25	14.809,36	115.188	421.483,83
2011	440.291,72	3.033,56	21.689,97	125.839	440.291,72
2012	439.173,84	2.797,49	28.402,56	76.736	439.173,84

*Includes household and public waste / Source: Goiânia 2012 - COMURG. Organised by the author.

According to the panorama of solid waste in Brazil developed by Abrelpe (2012), the rate of generation of solid urban waste in the state of Goiás is 1.050 kg/inhabitant/day (this figure does not include construction waste and other waste). Adopting this waste generation figure for the population of Goiânia, there would be 1,367.10 tonnes/day and 41,013.03 tonnes/month of urban waste generated. It is known that this rate is estimated and that the amount of waste generated may be higher than 41 tonnes/month, however, the monthly collection average published by COMURG is 35,000 tonnes. Table 11 shows the destination of the waste collected in Goiânia by COMURG.

Table 11 - Destination of solid waste collected in the municipality of Goiânia - 2011

Type of destination	Mass (tonne)	%
Recycled	1.700	4
Incinerated	250	1
Open burning	0	0
Disposed of in a "rubbish dump"	0	0
Disposed of in landfill	33.000	95
Total solid waste	35.000	100

Source: Goiânia 2012 - COMURG / Collection Directorate

Source note: The quantities of waste indicated represent the average of what is collected each month. In addition to the 35,000 tonnes per month, there is also 70,000 tonnes per month of removal waste (construction waste + cleaning waste).

There is a divergence between the way the figures were presented by COMURG in Table 10 and those in Table 11. In the former, the amount of recyclable waste that is collected is not included in the total amount of solid waste collected, so much so that the first column is the same as the last, which shows the amount of waste that is disposed of in landfill. In table 11, recycled waste makes up 4% of the total collected each month (35,000 tonnes). In this case, if part of the waste collected is destined for recycling, then in table 10, the total amount of waste that is disposed of in landfill should be less and not equal to the total collected.The collection and disposal of solid waste is fundamental to maintaining public health. The accumulation of rubbish in public places and/or vacant lots attracts insects, venomous animals and rodents that cause various diseases such as dengue, chikungunya, zika, yellow fever, leishmaniasis, accidents with venomous animals and others. In addition to health problems, the disposal of waste in inappropriate places such as rubbish dumps, vacant lots or even in streams and/or rivers causes environmental problems such as pollution of the soil and surface and groundwater, which directly or indirectly affects the health of the population.

3.3 Impacts of deficiencies in basic sanitation services in Goiânia-GO on the health of the population

Sanitation as an instrument for promoting health still needs to overcome numerous political, economic and cultural obstacles, among others, in order to reach the population living in urban and rural areas, including small municipalities. Most of the health problems affecting the world's population are in no way related to the environment (HELLER, 1997).

Access to basic sanitation services is one of the factors that promotes quality of life, well-being and environmental health. In Brazil, diseases caused by poor or inadequate sanitation, especially in poorer regions, have worsened the country's epidemiological situation. Diseases such as dengue, cholera, schistosomiasis, leptospirosis and diarrhoea, among others, are examples of this (BRASIL, 2007).

> The illnesses associated with poor or non-existent environmental sanitation and the consequent improvement in health due to the implementation of such measures have been the subject of discussion in various studies. Among these diseases, diarrhoea and parasitic diseases, particularly worms, have received attention from scholars and health authorities around the world (MORAES; 199, p.282).

A large number of diseases are related to the improper disposal of human waste, namely: hookworm disease, ascariasis, amoebiasis, cholera, infectious diarrhoea, bacillary dysentery, schistosomiasis, strongyloidiasis, typhoid fever, paratyphoid fever, salmonellosis, teniasis and cysticercosis. The modes of transmission of these diseases occur in various ways:

1) Direct skin contact with soil contaminated with helminth larvae from the faeces of people with parasitic diseases. When they come into contact with the skin, the parasite larvae settle in the human intestine;

2) Direct skin contact with water contaminated by cercariae. That's why it's not advisable to bathe in streams and lakes, especially in unknown regions or those where there are confirmed cases of schistosomiasis;

3) By ingesting food and water contaminated by faeces. This is how ascariasis, amoebiasis, typhoid and paratyphoid fevers, among others, are transmitted;

4) Through the bite of insect vectors or eating food contaminated by vectors such as flies;

5) By eating food contaminated by contact with the human hand. This means of transmission is closely related to a lack of hygiene and is considered the main mode of transmission of infectious diarrhoea, conjunctivitis and mycoses (BRASIL, 2007; HELLER, 1997; TEIXEIRA, HELLER, 2005).

In this research, diseases related to the absence or inefficiency of basic sanitation were divided into five categories. The diseases belonging to each category were selected from the data available on DATASUS. In the category of faecal-oral transmission diseases, it was decided to include data on "other intestinal infectious diseases" because they are closely related to parasitic diseases that are not specified in DATASUS.

Despite the structure of basic sanitation services in Goiânia, the population is still affected by various illnesses that are related to the absence or inefficiency of these services (Table 12). There were more than 600 hospitalisations for amoebiasis and viral hepatitis (except type B), and a total of five and 48 deaths, respectively, from 2008 to 2016 in Goiânia. With regard to confirmed and reported cases of viral hepatitis (including type B), DATASUS recorded 2,831 cases in Goiânia between 2007 and 2015 (BRASIL, 2015).

Table 12 - SUS hospital morbidity by disease category: number of hospital admissions, total value of public spending and number of deaths, January 2008 to November 2016, Goiânia-GO

Category	Diseases	No. of	Total value (R$)	N° Deaths

		hospitalisations		
Faeco-oral transmission diseases	Amebiasis	683	251.051,58	5
	Diarrhoea	14.103	4.649.993,59	61
	Other intestinal infectious diseases	18.322	6.761.947,03	91
	Cholera	761	335.852,06	14
	Typhoid and paratyphoid fever	136	43.976,23	0
	Polio sequelae	83	65.066,13	0
	Viral hepatitis[1]	636	310.897,54	48
Diseases transmitted by insect vectors	Haemorrhagic fever due to dengue virus	1.870	979.556,54	59
	Classic Dengue	18.865	5.815.237,74	78
	Yellow fever	3	1.439,48	-
	Filariasis	58	30.938,13	-
	Leishmaniasis	326	400.919,56	35
Diseases transmitted by contact with contaminated water	Schistosomiasis	3	1.252,25	-
	Leptospirosis	34	47.745,27	2
Hygiene-related illnesses	EYE DISEASE			
	T racoma	0	-	0
	Conjunctivitis	274	52.877,13	0
	SKIN DISEASES			
	Mycoses	1.042	1.019.451,31	61

[1] Does not include Hepatitis B
Source: BRASIL (2016), Ministry of Health-DATASUS.

The number of hospital admissions due to diarrhoea and other intestinal infectious diseases totalled more than 32,000 in Goiânia in the period evaluated. Total public spending on these hospitalisations was more than 11 million reais. With regard to polio, national vaccination campaigns against the disease began in 1979 and the last cases of polio in Brazil were recorded in 1989 (BRASIL, 2003).
That's why the 83 hospitalisations recorded were due to the consequences of polio.

Typhoid fever and cholera are diseases that are associated with low socioeconomic levels, as well as areas where sanitation, personal hygiene and environmental conditions are precarious. "In Brazil, typhoid fever occurs in endemic form, with overlapping epidemics, especially in the North and Northeast regions, reflecting the living conditions of their populations." (PORTAL DA SAÚDE SUS, 2014). The disease can also occur in more developed areas of the country, especially on the outskirts of large urban centres. Between 2008 and 2016, 761 hospitalisations for cholera, 14 deaths and 136 cases of typhoid fever were recorded in Goiânia.

Although there is a vaccine against cholera and typhoid fever, the Ministry of Health's National Immunisation Programme (PNI) does not recommend vaccination against this disease, as the vaccines have low efficacy and a short duration of immunity. Vaccination is recommended by the World Health Organisation (WHO) for travellers to areas where cases of the disease occur or for people living in areas of high endemicity (PORTAL DA SAÚDE SUS, 2015). In addition to knowing the number of hospitalisations due to diseases, it is interesting to know the most vulnerable age group. Figure 18 shows the occurrence of hospitalisations for diseases belonging to the faeco-oral transmission category by age group.

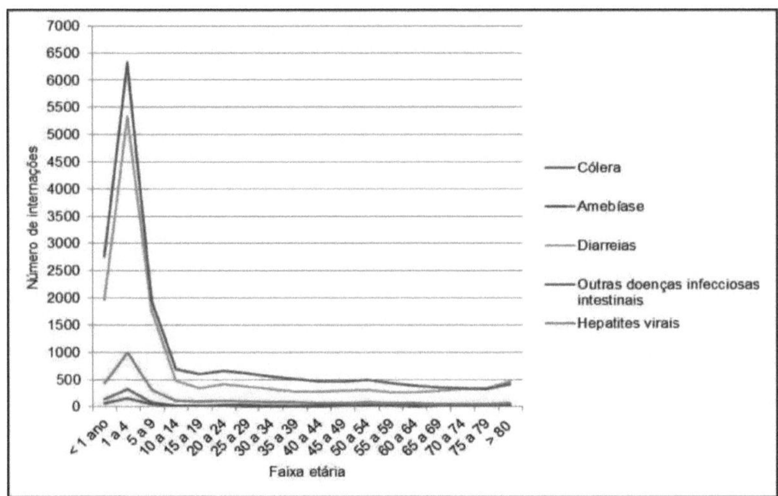

Figure 18 - Number of hospitalisations by age group according to the category of faecal-oral transmission diseases, Goiânia-GO from 2008 to 2016
Source: Ministry of Health - DATASUS, 2016.

Of the five diseases shown in Figure 18, all except viral hepatitis had a higher incidence in the 0-4 age group. Of the relative total number of hospitalisations for amoebiasis, cholera, diarrhoea and other infectious intestinal diseases, 32%, 61%, 52% and 50%, respectively, were children under one year old and up to four years old. Childhood diarrhoea is one of the main diseases related to poor sanitation services. It is estimated that the mortality of children under five years of age reaches 1.5 million per year worldwide (BLACK et al., 2003).

With regard to diseases belonging to the category transmitted by insect vectors, dengue fever and haemorrhagic fever due to the dengue virus stand out. There were 18,865 hospitalisations due to dengue, more than five million reais in public spending and 75 deaths recorded between 2008 and November 2016; hospitalisations due to haemorrhagic fever were almost 2,000 and 59 deaths. With regard to confirmed and notified cases of dengue in Goiânia, DATASUS recorded 116,459 cases between 2007 and 2012 (BRASIL, 2012). There were three hospitalisations for yellow fever, 58 for filariasis and 326 for leishmaniasis (Table 12). In addition to hospitalisations, DATASUS recorded 648 confirmed and notified cases of leishmaniasis in the municipality of Goiânia between 2007 and 2015 (BRASIL, 2015). Figure 19 shows the distribution of hospitalisations for this category of disease by age group.

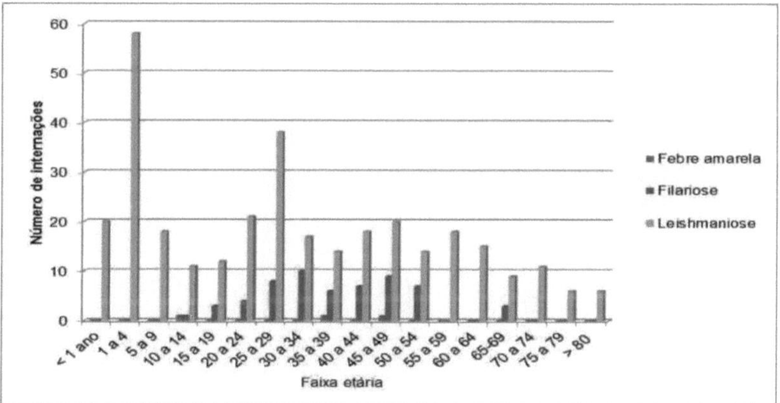

Figure 19 - Number of hospitalisations by age group according to the category of diseases transmitted by insect vectors, Goiânia-GO from 2008 to 2016
Source: Ministry of Health-DATASUS, 2016.

Leishmaniasis, which is transmitted by the bite of insects of the *Lutzomyia* and *Psychodopigus* genera, was the disease that affected all age groups, with the highest number of hospitalisations in absolute (78) and relative (24%) figures being in the zero to four year-old age group. Figure 20 shows the distribution of dengue and dengue haemorrhagic fever hospitalisations by age group.

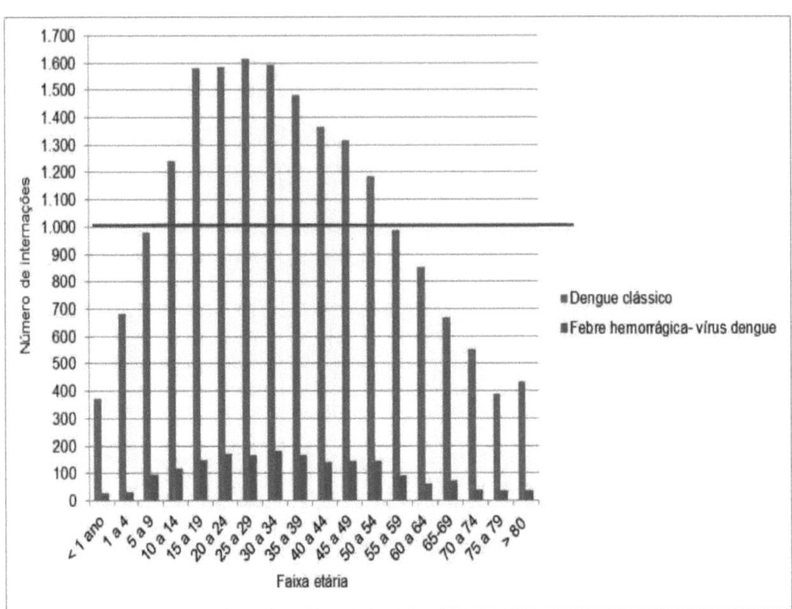

Figure 20 - Number of hospitalisations by age group according to the category of vector-borne diseases (Dengue and Haemorrhagic Fever), Goiânia-GO from 2008 to 2016. Source: Ministry of Health-DATASUS, 2016.

Dengue fever and haemorrhagic fever affected all age groups. The number of hospitalisations due to dengue reached over 1,000 in the 10 to 54 age group. In relative numbers, the distribution of dengue hospitalisations was as follows: 6% in the zero to four age group, 12% in the five to fourteen age group, 67% of hospitalisations in the 15 to 59 age group and 15% in the elderly age group (over 60).In addition to the data on hospitalisations, it is important to know the number of reported dengue cases in Goiânia. Data released by Goiânia City Hall through the Municipal Health Department shows that the number of reported dengue cases in the municipality is high. By the second week of December 2016, more than 62,000 suspected dengue cases had been reported, an incidence of 4,429.5 cases per 100,000 inhabitants. The serotypes identified were DEN-1 (70%) and DEN-4 (30%) (GOIÂNIA, 2016). Figure 21 shows the evolution of the number of dengue cases reported per year in Goiânia between 2003 and 2016.

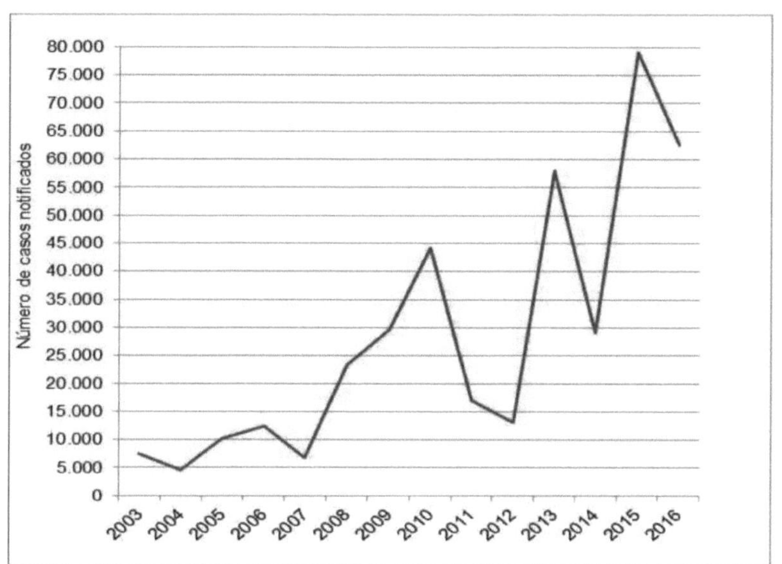

Figure 21 - Dengue cases reported in Goiânia between 2003 and 2016 Source: Goiânia (2016).

Figure 21 shows that the number of dengue notifications in Goiânia rose uninterruptedly between 2007 and 2010. There was a significant drop in 2012, but the following year notifications rose again. The highest number of dengue cases (79,095) was recorded in 2015 during the period studied. The number of dengue deaths also showed periods of high and low in the same period (Figure 22).

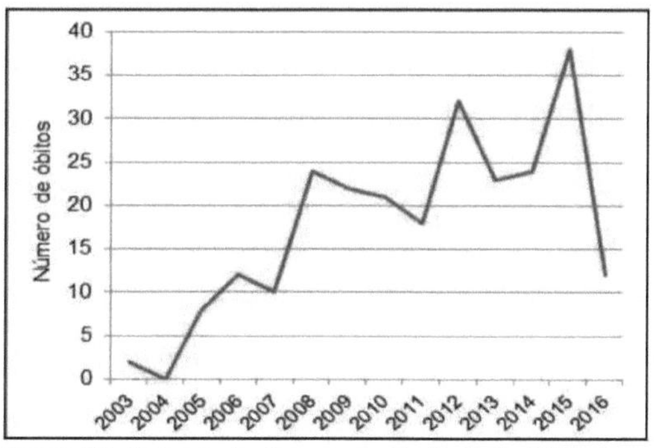

Figure 22 - Dengue deaths reported in Goiânia-GO between 2002 and 2016 Source: Goiânia (2016).

According to municipal data, 2012 and 2015 saw the highest number of deaths from dengue, 32 and 38, respectively. Data from the Ministry of Health - DATASUS (2016) recorded 78 deaths from dengue in Goiânia between January 2008 and November

2016, of which two were children aged between one and nine, 43 were people aged between 15 and 59 and 32 were adults aged over 60 (BRASIL, 2016).

In 2014 and 2015, two new diseases were discovered, Chikungunya Fever and Zika Virus, which are transmitted by the same insect that transmits dengue fever and yellow fever, the *Aedes Aegypty.* In 2015, 42 probable cases of acute Zika virus disease and 10 probable pregnant women were reported. In 2016 the number of probable cases rose to 8,863 and 489 pregnant women. Figure 23 shows the cases of Chikungunya fever recorded between 2014 and 2016.

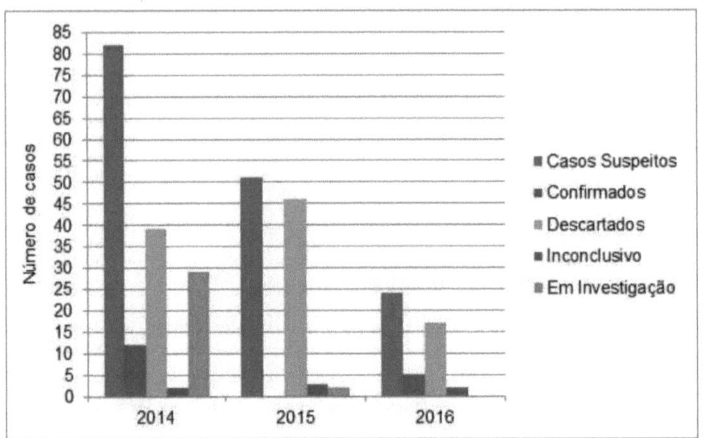

Figure 23 - Notified cases of Chikungunya fever in residents of Goiânia, 2015-2016
Source: SINAN/GDAT/DVE/SVS/SMS - Goiânia (2016), prepared by the authors.

Leptospirosis is a disease caused by a bacterium (Leptosp/ra *interrogans)* found in rat urine, which normally spreads through floodwater and sewage. Transmission of the disease is closely associated with poor solid waste management and urban drainage services. Poor solid waste management becomes a health problem because it favours the proliferation of vectors and rodents that cause diseases such as infectious diarrhoea, amoebiasis, salmonellosis, helminthiasis such as ascariasis, teniasis and other parasites. "It also serves as a breeding ground and hiding place for rats, animals that are involved in the transmission of bubonic plague and leptospirosis" (BRASIL, 2007, p. 287). Schistosomiasis is a parasitic disease also acquired through human contact with contaminated water:

> Schistosomiasis is a transmissible, parasitic disease caused by trematode worms of the genus Schistosoma. There are currently six species of Schistosoma (S.

mansoni, S. haematobium, S. japonicum, S. intercalatum, S. mekongi and S. malayensis) that can cause disease in humans, and on the American continent there is only S. mansoni. In Brazil, schistosomiasis mansoni is endemic in a vast area of the country and is still considered a serious public health problem because it affects millions of people, causing a significant number of serious illnesses and deaths every year (BRASIL, 2008, p. 11).

In Goiânia, there were three hospitalisations for schistosomiasis and 34 for leptospirosis. The highest number of hospitalisations for leptospirosis was among the 25 to 34 and 50 to 54 age groups (Figure 24).

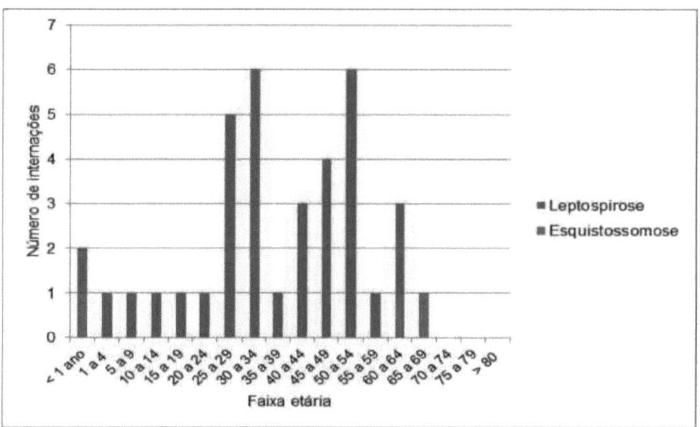

Figure 24 - Number of hospitalisations by age group according to the category of diseases transmitted by contact with water, Goiânia-GO from 2008 to 2016

Source: Ministry of Health-DATASUS, 2016.

Figure 25 shows the number of hospitalisations by age group in the hygiene-related illnesses category.

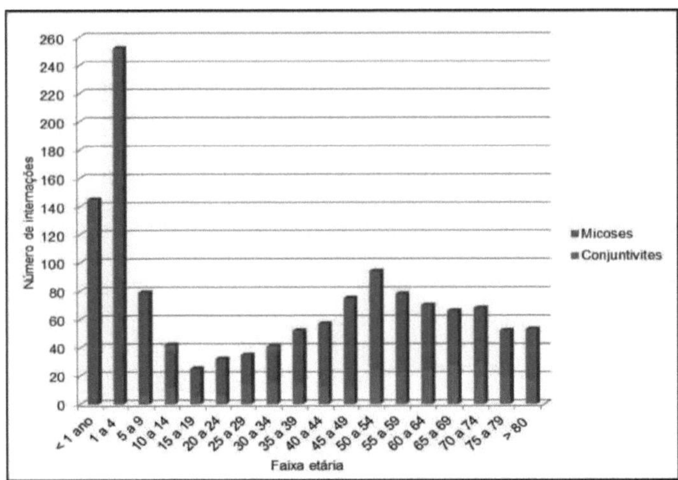

Figure 25 - Number of hospitalisations by age group according to the category of hygiene-related diseases, Goiânia-GO from 2008 to 2016 Source: Ministério da Saúde-DATASUS, 2016.

With regard to the category of hygiene-related diseases, conjunctivitis and mycosis stand out, accounting for 274 and 1,042 cases, respectively, of hospitalisations in Goiânia between 2008 and November 2016. The distribution of these diseases affected all age groups, with 38 per cent of hospitalisations for mycoses among children aged zero to four.

CHAPTER 4

BASIC SANITATION IN GOIÂNIA - GO: A CASE STUDY IN THE JARDIM CERRADO NEIGHBOURHOOD

Jardim do Cerrado is located in the western region of Goiânia, close to the border with the municipality of Trindade. It covers an area of around 500 hectares and is 25 km from the city centre. Paved access to the neighbourhood is via GO-060 and Avenida Luísa M. Coimbra Bueno; there is an unpaved access via Avenida Vinícus de Moraes, which connects Jardim do Cerrado to Conjunto Vera Cruz. There are important permanent preservation areas in the area, namely: the source of the Cruz stream and the source of the tributary of the Cruz stream, which has an unknown name, see Figure 26.

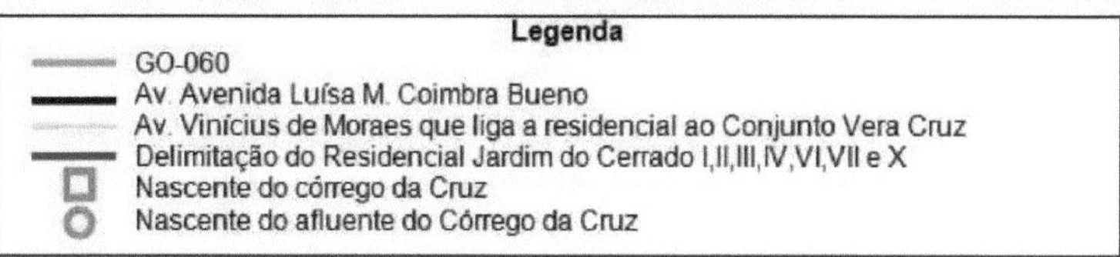

Figure 26 - Accessibility to Residencial Jardim do Cerrado and permanent preservation areas in the neighbourhood
Source: Google Earth (2016), modified by the authors.

According to a study by Lucas (2016), the Jardim do Cerrado residential development is one of the largest projects financed by the Minha Casa Minha Vida

Programme (PMCMV) in the state of Goiás. Ten thousand housing units are to be built to cater for a population of 40 thousand. Construction was scheduled to be completed in 2016, but there are still some modules that have not yet been built. The residential development was divided into 11 modules, with modules I, II, III, IV, VI, VII and X already consolidated, as shown in Figure 27.

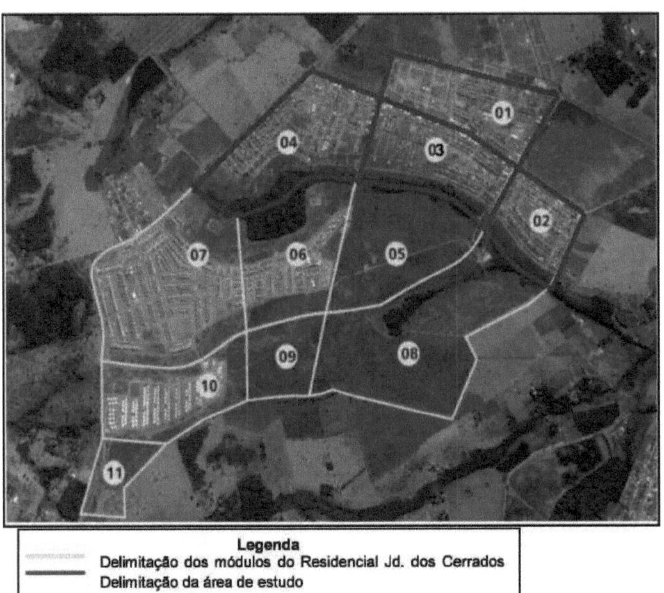

Figure 27 - Aerial image of Residencial Jardim dos Cerrados and the divisions of the stages completed and to be built
Source: LUCAS (2016, p.132). Adapted by the authors.

Modules I to IV are intended for residents of at-risk areas and less socially favoured families in income bracket 1 (up to R$1,600.00) according to the PMCMV classification; the type of housing in this bracket is a detached house on the plot. In modules VI and VII there are housing units for both families in income bracket 1 and those with higher incomes, which are classified as bracket 2 (up to R$3,275) and bracket 3 (up to R$5,000); the type of housing units in these modules are condominiums of houses, overlapping houses and townhouses (LUCAS, 2016). The way in which homes are bought and paid for according to income bracket is also different, as described by Lucas (2016):

> Jardim do Cerrado residential housing is being built in bands 1, 2 and 3. For band 1, resources from the Residential Lease Fund (FAR), which comes from

> the general budget of the Federal Government, are mainly used to make it possible to build housing units. In this modality, the municipality is responsible for actions that facilitate the realisation of the projects, in particular the indication of priority areas to build units, contribution of resources, exemption from taxes, indication of residents and execution of technical and social work. In this group, which is more socially fragile, most of the subsidy comes from the Federal Government. The beneficiary pays 5% of their monthly income, with a minimum instalment of R$25.00.
> In bands 2 and 3, the MCMV grants direct financing to individuals, formalised through a partnership with an organising body or company. The company makes the necessary choices regarding location, density, construction technology and other points. The unit is sold on the formal market and the buyer, if they have a family income of up to R$5,000, can get financing and a subsidy to buy it. Thus, in bands 2 and 3, subsidies are granted in the form of a discount to reduce the value of the instalments or payment of part of the purchase or construction of the property. The discount is calculated based on the applicants' gross monthly family income, taking into account the location of the property, i.e. the municipalities in the national territory and the operational modalities (LUCAS, 2016, p.128).

The difference is not just in the way the houses are bought and paid for. It's in the type of homes, the finishes, the neighbourhood's urban infrastructure and sanitation. It's worth noting that despite the differences between the modules, both are located on the outskirts of the city and are socio-spatially segregated. These differences will be described in more detail below.

4.1 Differences between the modules at Residencial Jardim Cerrado

Considering that the first four modules of Residencial Jardim Cerrado were intended for income bracket one (poorer families) and modules six and seven for income brackets two and three of the PMCMV. The first distinction between these modules refers to the type of homes (Figures 28 and 29).

Figure 28 - Distinction between property types in Jardim Cerrado I to IV (a) and Jardim Cerrado VI and VII (b). Source: Personal collection. Date: 17 December 2016.

In modules one to four, the houses were built on individual plots. The house is 40.8 metres long[2] and consists of two bedrooms, a bathroom and a living room with a combined kitchen. In the other modules, the residences are made up of detached houses on the plot, overlapping houses and townhouses. The two-storey buildings are made up of eight housing units measuring 42m² each, with each unit comprising two bedrooms, a living room, kitchen, bathroom and laundry area (LUCAS, 2016) (Figure 29).

Figure 29 - (a) Model home on individual plots in modules I to IV and (b) townhouses in gated communities in modules VI and VII Source: Personal collection. Date: 17 December 2016.

The difference between the finishes, organisation and architecture of the houses is striking. In image "a", the houses are simple, with no pavements and no sense of beauty or well-being. In image "b", the houses have a better finish, modern paintwork, a garden and an electric fence. The architecture of the condominium conveys a good impression and well-being to the residents. Figures 30 show the pavements and streets of the residential modules.

Figure 30- Difference between the pavements and streets of the Jardim Cerrado modules: (a) and (b) Modules for income group one and (b) modules for income groups two and three.
Source: Personal collection. Date: 17 December 2016.

All the roads in Residencial Jardim Cerrado VI and VII have kerbs and cemented and grassed pavements for pedestrians to cross. In the first modules (image "a" in Figure 30), most of the roads are unpaved and those that are paved do not have curbs or structured pavements. Some pavements are more than 70 centimetres uneven (Figure 31), which makes it difficult for pedestrians to walk, especially the elderly and people with reduced mobility.

Figura 31 - Unevenness of the pavement in relation to the street in Residencial Jardim Cerrado I to IV.
Source: Personal collection. Date: 17 December 2016.

With regard to leisure, the residents of both modules don't have many options, given the great distance between the neighbourhood and the city centre. Even so, it is possible to see some distinctions, which are shown in Figure 32.

Figura 32 - (a) Football pitch for sports in Jardim Cerrado I to IV and (b) small sports court inside the condominium in Jardim Cerrado VI and VII.
Source: Personal collection. Date: 17 December 2016.

The type of sewage disposal is also offered differently to the residents of Jardim Cerrado. For the residents of the first modules, sewage is disposed of via septic tanks and drains, while for the others, household effluent is disposed of via the general sewage network and taken to SANEAGO's sewage treatment plant, where after undergoing the appropriate treatment, it is taken through pipes to the receiving body (Figure 33).

Figure 33 - Types of sewage disposal: a) Septic tank and drain in modules I to IV and b) Sewage treatment plant to serve modules VI and VIL Source: Personal collection. Date: 17 December 2016.

Septic tanks and cesspools are viable ways of disposing of sanitary sewage when done correctly. In addition, septic tanks are designed based on the number of inhabitants in each household. If this number is exceeded, the capacity of the tank can be exceeded and the frequency of maintenance reduced. This type of problem no longer exists with the general sewage collection system.

4.2 Results of the survey with the residents of Jardim Cerrado modules I to IV

Although Residencial Jardim Cerrado is made up of several modules, the opinion poll was restricted to the first four modules because they are homes for low-income families. The aim of this survey was to ascertain residents' opinions on issues related to basic sanitation (water supply, sewage, solid waste and rainwater drainage) in order to better understand the reality of the basic sanitation infrastructure in the neighbourhood and whether or not it has been provided satisfactorily to residents. The questionnaire was divided into five parts: 1) general questions to identify the population's level of knowledge and satisfaction about basic sanitation; 2) questions about water supply; 3) questions about sanitation; 3) questions related to solid waste; 4) questions about rainwater drainage; 5) questions about diseases related to the lack or inefficiency of basic sanitation.

The inclusion of general questions about basic sanitation was important for identifying the population's level of knowledge on the subject. Once these people had been identified, it was possible to explain the subject to them. Table 14 and Figure 34 show the residents' answers to general questions related to basic sanitation.

Table 13 - Knowledge of Jardim Cerrado residents about basic sanitation, Goiânia-GO.

QUESTIONS	ANSWERS			
	Yes	%	No	%
Do you know what basic sanitation is?	37	61,7	23	38,3
In your opinion, can basic sanitation interfere with people's health?	48	80	12	20

Source: Prepared by the authors.

Although more than half of the interviewees answered that they knew what basic sanitation meant, many were still confused when asked to tick which services were part of basic sanitation (Figure 34).

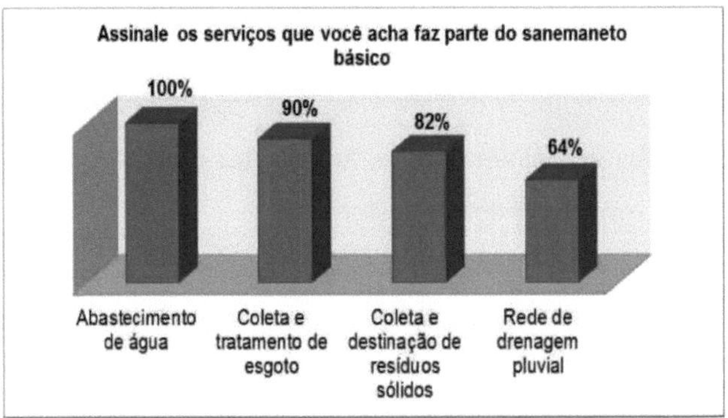

Figure 34 - Opinion of Jardim Cerrado residents on basic sanitation services, Goiânia-GO.
Source: Prepared by the authors.

The rainwater drainage system and the collection and disposal of solid waste were the services that most left the residents who took part in this interview in doubt. Some didn't know what rainwater drainage meant and had to explain it beforehand. There is a lack of clarification on the subject and the population is still confused about which services belong to basic sanitation. After explaining basic sanitation services in detail, the survey participants answered whether or not they were satisfied with the sanitation in their neighbourhood, see Figure 35.

Figure 35- Opinion of Jardim Cerrado residents on basic sanitation services.
Source: Prepared by the authors.

While 55% of the participants were dissatisfied with the basic sanitation services offered in the neighbourhood, 45% were satisfied. The most praised service was water supply, while the most criticised was sanitation.

a) Water supply

With regard to water supply, the households in Residencial Jardim do Cerrado are 100 per cent served by SANEAGO's general network. The treated water comes from the Meia Ponte system and has to be pumped several times before it reaches the homes (Figure 36).

Figure 36 - SANEAGO reservoirs that distribute water to the residents of Jardim Cerrado and other nearby

neighbourhoods, Goiânia-GO.
Source: Personal collection, Date: 19/11/2016 and 17/12/2016.

Until January 2017, SANEAGO has been able to supply water in quantity and quality to the residents of Jardim do Cerrado, without interruptions in supply or the need to expand the network. However, as the population is projected to grow and densify, it is possible that in the coming years the concessionaire will need to invest in expanding the network in order to meet the demand for water (verbal information)[5] . Figure 37 shows the residents' response to the type of water supply and quality.

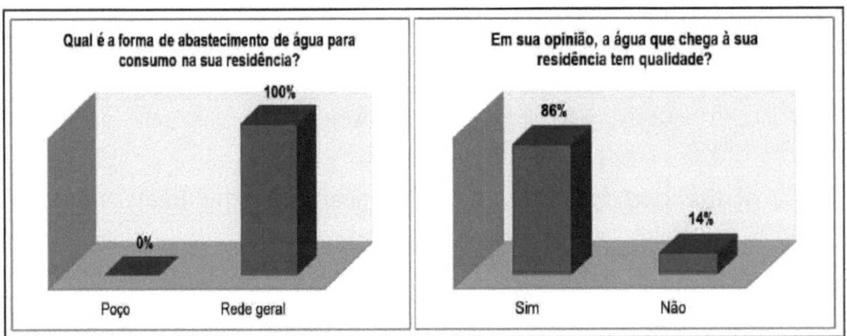

Figure 37 - (a) Type of water supply and (b) water quality according to information from residents of Jardim Cerrado, Goiânia-GO.
Source: Prepared by the authors.

All the homes of the residents who took part in the survey are supplied by the general water network, which confirms what Saneago's water manager said about the universalisation of the water supply in this residential area. With regard to the quality of the water, only 32 per cent of the residents think that the water that reaches their home is of poor quality. Figure 38 shows the residents' response to the lack of water.

[5] Information provided by SANEAGO's Water Manager, Engª Lúcia, in Goiânia- GO, in October 2016.

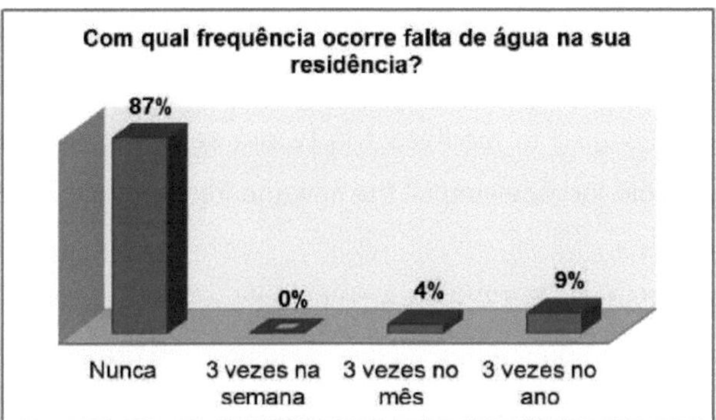

Figure 38 - Residents' opinion on the frequency of water shortages in Jardim Cerrado, Goiânia-GO.
Source: Prepared by the authors.

When asked about the frequency of water shortages, 87% of the participating residents said that there is never a water shortage in the neighbourhood, while 13% said that there is a water shortage in the neighbourhood. Of this total, 3% said that water shortages only happen about three times a month and 10% said that they happen three times a year.

Chapter two showed that more than 30 million Brazilians do not have access to a quality drinking water supply. Medeiros Filho (2008) points out that when the population density in a given community increases, the most viable basic sanitation solution is the implementation of a public water supply system. This solution becomes the most suitable because it is more efficient in preserving water sources and controlling the quality of the water distributed to the population. Xavier and Loreto (2016) emphasise that in order to consider water supply efficient, one must consider water supply in quantity and quality. Sufficient quantity to meet all the consumption needs of a given community and adequate quality for all consumption purposes.

b) Sanitary sewerage

"The sanitary sewerage system is the set of works and facilities that provides collection, transport and removal, treatment, and final disposal of wastewater, in an appropriate manner from a sanitary and environmental point of view" (XAVIER; LORETO, 2016, p. 3).

The sanitary sewage system exists to prevent human waste from coming into contact with the population, food and water sources, in order to reduce or even eradicate disease-transmitting vectors (RIBEIRO and ROOKE, 2010).The type of sanitary sewage in the homes belonging to modules I to IV are septic tanks and drains. The septic tank is responsible for receiving all the sewage from the home.

According to Brazilian Standard NBR 7229/93, which deals with septic tanks, the distance between the tank and the boundaries of land, buildings and drains is at least 1.5 metres. One of the reasons for this is to prevent damage to nearby buildings, the drain and neighbours in the event of a leak.

However, what we see in reality is non-compliance with this standard: the septic tanks in Jardim do Cerrado homes are practically leaning against walls, houses and drains, as shown in Figure 39.

Figure 39 - Septic tanks and drains in Jardim do Cerrado homes built irregularly.
Source: Personal collection, date: 19/11/2016.

In image (a) it can be seen that the septic tank is very close to the drain and the house, and in the other image (b), the tank is practically against the wall. In both examples, the minimum distance of 1.5 metres between the septic tank and other buildings - established by the NBR - was not complied with. In Jardim do Cerrado I, sewage was found running out into the open on the neighbourhood streets - the stench was very strong (Figure 40).

Figura 40 - Domestic sewage overflowing from a house and running down the neighbourhood streets. Source: Personal collection, date: 19/11/2016.

Residents were very dissatisfied with the type of sewage system in their neighbourhood. The answers are shown in Table 15.

Table 14 - Answers from Jardim Cerrado residents on issues related to sanitation.

QUESTIONS	Yes	No
Did you know that there is a WWTP in your neighbourhood?	80%	20%
Are you satisfied with your home's sewage system (septic tank and drain)?	26%	74%
Have you received any advice on the operation and maintenance of the septic tank?	4%	96%

Source: Prepared by the authors.

All the residents who took part in the survey said when the questionnaire was administered that the tanks give off a bad odour (the smell comes back through the drains in the toilets), that there are lots of cockroaches that are attracted to the tank and that the tank is very expensive to maintain. They all complained about the financial costs cleaning the septic tank, because it can't cope with the capacity of the effluent that is discharged. In homes where up to four people live, they clean the septic tank every six months; above this number, residents reported that it should be cleaned every four months. The companies that provide this type of service charge between R$200.00 and R$250.00 to clean a cesspit.

Faced with the problems and damage caused by the septic tank, the residents questioned why they don't have a general sewage system when there is a sewage treatment plant in the neighbourhood. They were also very unhappy about having to put up with the stench coming from the sewage treatment plant every afternoon, and also the stench from the septic tanks, where there are certainly technical problems that don't eliminate the odour.

It is known that in Brazil, the deficit in water supply and sewage services is more pronounced in low-income populations, and that the greatest public health problems are found in this low-income population group (SILVA; MACHADO, 2001). These shortcomings in sanitation show the characteristics of inequalities in terms of region, family income and location of the home (HELLER; NASCIMENTO, 2005). It is clear that the inequalities that exist in Brazil are not only expressed from an economic point of view, but also in terms of basic sanitation. Jardim do Cerrado is a good example of this.

The other modules in Jardim do Cerrado, which are predominantly intended for families in bands 2 and 3 of the PMCMV, have different sewage systems. A sewage treatment plant was built downstream of modules V to XI, which receives all the sewage produced (Figure 41).

Jardim do Cerrado: modules I to IV with septic tank and drainage system. **Detail 1-** Location of the Jardim dos Cerrados Sewage Treatment Plant operated by SANEAGO. Next to the sewage treatment plant there is a reservoir, created by the waters of the tributary of the Córrego da Cruz, used for recreation by local residents.

Figura 41 - Location of the Jardim dos Cerrados WWTP operated by SANEAGO, Goiânia-GO. Source: Google Earth, 2016.

The WWTP has a treatment capacity of 2,300 l/s of domestic sewage. It currently receives and treats sewage from modules VI and VII of Jardim dos Cerrados and has not yet exhausted its treatment capacity (verbal information)[6] . Although SANEAGO does not provide information on the type of treatment used at this WWTP and the efficiency of Biochemical Oxygen Demand (BOD) removal, it should be pointed out that there are rules on the conditions and standards for discharging sanitary effluents into water bodies, such as CONAMA Resolution No. 357 of 2005, amended by CONAMA Resolution No. 430 of 2011:

> Art. 21: For the direct discharge of effluents from sanitary sewage treatment systems, the following specific conditions and standards must be complied with: I - Effluent discharge conditions: a) pH between 5 and 9; b) temperature: less than 40°C, and the temperature variation of the receiving body must not exceed 3°C at the limit of the mixing zone; c) sedimentable materials: up to 1 mL/L in a 1-hour Inmhoff cone test. For discharge into lakes and ponds, where the circulation speed is practically zero, sedimentable materials must be virtually absent; d) Biochemical Oxygen Demand - BOD 5 days, 20°C: maximum 120 mg/L, and this limit may only be exceeded in the case of effluent from a treatment system with a minimum BOD removal efficiency of 60%, or by means of a self-depuration study of the water body that proves compliance with the targets of the receiving body's framework. e) hexane-soluble substances (oils and greases) up to 100 mg/L; and f) absence of floating materials (BRASIL, 2011, Art. 21).

The reservoir located next to the WWTP was created by the waters of the tributary of the Cruz stream. As leisure options and community facilities are precarious in the area, this reservoir is used by local residents for leisure. Every Sunday a large number of families, young people and children gather here to bathe, play, chat and have fun; the reservoir has become a meeting and leisure point for

[6] Information provided by the technical operator of the Jardim dos Cerrados WWTP in Goiânia, date: 21 November 2016.

people (Figure 42).

Figure 42 - Dam next to the WWTP that has become a leisure area for Jardim do Cerrado residents. Source: Personal collection, date: 21/11/2016.

The reservoir used for primary recreation by residents is located in a pleasant spot, next to paths covered in buritis. As residents like to go there to bathe, barbecue and listen to music, it would be an ideal place to build a park with an area reserved for bathers and *camping.* However, the stench coming from the sewage treatment plant, which was built next door, has degraded the environment. In addition, another worrying factor that could have a negative impact on the health of the people who frequent the area is the issue of water quality and the excess of solid waste disposed of improperly (Figure 43).

Figure 43 - Environmental conditions of the reservoir and riparian forest: (a) and (b) physical characteristics of the water (colour and odour) of suspect quality; (c) and (d) solid waste improperly disposed of in the reservoir's permanent preservation area.
Source: Personal collection, date: 21/11/2016.

The images in Figure 43 reveal the carelessness of the resident population with the permanent preservation area (APP), by improperly disposing of waste in the water resource and riparian forest. On the other hand, there is also a lack of interest on the part of the state in looking after this PPA area.

If the water in this reservoir, which is used for bathing and fishing by local residents, does not comply with the quality standards established by environmental legislation for this use, the population, especially children, are subject to various waterborne diseases, described in chapter two.

CONAMA Resolution No. 357 of 2005, which deals with the classification of bodies of water, states that class one and two fresh waters can be used "for primary contact recreation, such as swimming, water skiing and diving" (BRASIL, 2005, Art. 4º). The same resolution also sets out the quality conditions and standards for these waters. Class one and two freshwaters, which can be used for primary recreation, must have the following conditions and standards according to articles 14 and 15 respectively:

> **Art. 14. Class 1 fresh waters shall comply with the following conditions and standards:** I - water quality conditions: a) no chronic toxic effect on organisms, in accordance with the criteria established by the competent environmental body, or, in its absence, by renowned national or international institutions, proven by carrying out a standardised ecotoxicological test or other scientifically recognised method, b) floating materials, including non-natural foams: virtually absent; c) oils and greases: virtually absent; d) substances that communicate taste or odour: virtually absent; 6 e) dyes from anthropogenic sources: virtually absent; f) objectionable solid waste: virtually absent; g) coliforms
> thermotolerant coliforms: for primary contact recreation, the bathing quality standards set out in CONAMA Resolution 274 of 2000 must be complied with. For all other uses, a limit of 200 thermotolerant coliforms per 100 millilitres should not be exceeded in 80% or more of at least 6 samples collected every two months over a one-year period. E. Coli may be determined instead of the thermotolerant coliform parameter in accordance with the limits established by the competent environmental body; h) BOD 5 days at 20°C up to 3 mg/L 02; i) DO, in any sample, not less than 6 mg/L 02; j) turbidity up to 40 nephelometric turbidity units (UNT); l) true colour: natural colour level of the body of water in mg Pt/L; and m) pH: 6.0 to 9.0.
> **Art. 15:** The conditions and standards of **class** 1 provided for in the previous article **shall apply to** class **2 fresh waters**, with the exception of the following: I - the presence of colourants from anthropogenic sources that are not removable by conventional coagulation, sedimentation and filtration processes shall not be permitted; II - thermotolerant coliforms: for primary contact recreation use, CONAMA Resolution 274 of 2000 shall be complied with. For other uses, a limit of 1,000 thermotolerant coliforms per 100 millilitres should not be exceeded in 80% or more of at least 6 (six) samples collected over a period of one year, on a bimonthly basis. E. coli may be determined instead of the thermotolerant coliform parameter in accordance with the limits set by the competent environmental body; III - true colour: up to 75 mg Pt/L; IV - turbidity: up to 100 UNT; V - BOD 5 days at 20°C up to 5 mg/L 02; VI - DO, in any sample, not less than 5 mg/L 02; VII - chlorophyll a: up to 30 pg/L; VIII - cyanobacteria density: up to 50000 cel/mL or 5 mm3 /L; and, 10 IX - total phosphorus: a) up to 0.030 mg/L, in lentic environments; and, b) up to 0.050 mg/L, in intermediate environments, with a residence time between 2 and 40 days, and direct tributaries of a lentic environment (BRASIL, 2005, Art. 14 e 15).

c) Solid waste

One of the biggest problems faced by post-modern society is the high consumption of products and services, which consequently generates an excess of solid waste that is often disposed of in inappropriate places in the environment. The growing volume of waste aggravates the problems related to population agglomeration in urban areas and consequently reduces or increases the cost of landfill sites (MARTINS et al., 2016).

Solid waste poses sanitary, economic and, above all, aesthetic problems. According to local residents, the rubbish collection truck passes through the neighbourhood three times a week (3ª, 5ª and Saturday). When asked about the selective collection truck, one resident said that at first, as soon as the neighbourhood was opened, the truck passed through regularly, but that after a while it stopped passing. Figure 44 shows the residents' responses about their knowledge of the Selective Collection programme in Goiânia and how they separate household waste.

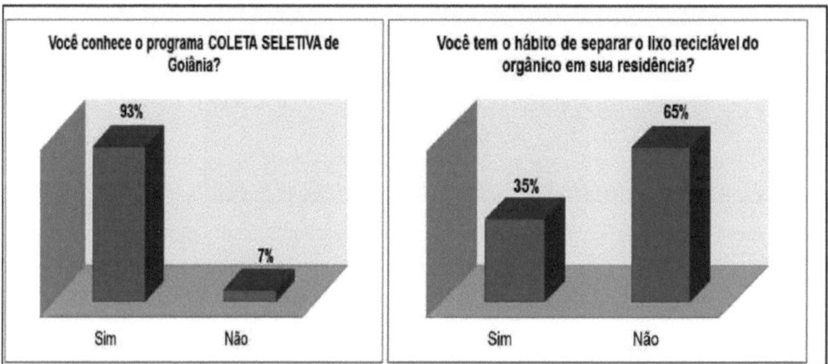

Figure 44 - Answers from Jardim Cerrado residents on issues related to household solid waste.
Source: Prepared by the authors.

Although 93% of the survey participants were aware of the selective collection programme, the percentage of those who don't have the habit of separating recyclable from non-recyclable waste is still high (65%). When you walk along the streets of Jardim do Cerrado I to IV, what you see most on street corners and vacant lots is solid waste, both domestic and construction waste and pruning debris (Figure 45).

Figura 45 - Solid waste in various parts of the Jardim do Cerrado Residential Centre. Source: Personal collection, date: 20/11/2016.

In some places, the amount of rubbish is so great that it even invades the streets and obstructs the passage of vehicles. The football pitch, for example, which should be a place for children and teenagers to play, has become a rubbish dump. As you walk around the neighbourhood, you can see that it is common for people to discard waste, especially construction waste, furniture, tree pruning and recyclable waste.

CONAMA Resolution 307 of 2002 establishes the following about construction waste:

> "Art. 4° Civil construction waste may not be disposed of in household waste landfills, in "dumping" areas, on slopes, bodies of water, vacant lots and in areas protected by law, subject to the deadlines defined in art. 13 of this Resolution (...) Art. 11. A maximum period of twelve months is established for municipalities and the Federal District to draw up their Integrated Plans for the Management of Civil Construction Waste, including Municipal Programmes for the Management of Civil Construction Waste from small volume generators, and a maximum period of eighteen months for their implementation (...) Art. 13. Within a maximum period of eighteen months, municipalities and the Federal District must cease the disposal of civil construction waste in household waste landfills and in "dump" areas. (BRAZIL, 2002, articles 4°, 11 and 13).

COMURG, the company responsible for urban cleaning in Goiânia, justifies the receipt of RCC in the municipal landfill, by means of COMURG Resolution No. 20 of 07 June 2016:

> Considering that Goiânia does not yet have an Integrated Construction Waste Management Plan and that the landfill uses RCC as a sub-base for the manoeuvring yards, the refurbishment of internal roads and as part of the cover layer for household waste, we believe that it is feasible to receive class A and C RCC (COMURG, 2016, p. 4).

COMURG, which also manages Goiânia's landfill, states in the same resolution that construction waste is not collected by COMURG, and that it must be taken to "the landfill by the generator itself or by private companies contracted by it, in skips or lorries" (COMURG, 2016, p. 5). However, the generator can only dispose of this waste by paying a fee; disposal is free for small generators (up to 500kg). Figure 46 shows residents' opinions on the difficulties in properly disposing of construction waste.

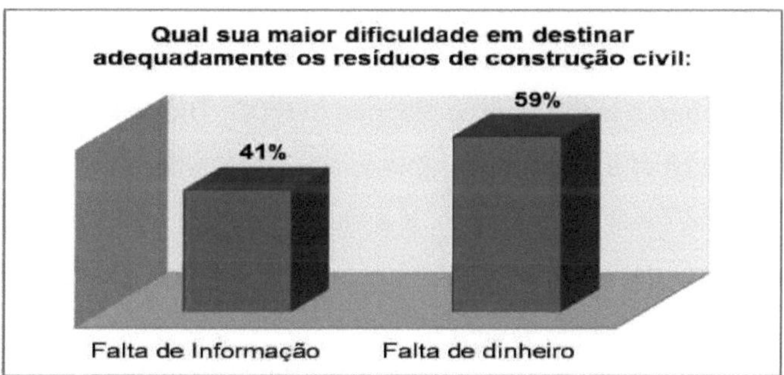

Figure 46 - Residents' opinions on the difficulties in disposing of CCW. Source: Prepared by the authors.

In the opinion of 42% of the participating residents, the biggest difficulty in properly disposing of their waste is a lack of information, i.e. they don't know what to do with it or where they should take it. Lack of financial resources is the biggest impediment in the opinion of 58 per cent; they know they should hire a skip, but don't have enough resources to do so. Figure 47 shows the guidelines on COMURG's website for disposing of rubble, furniture and household appliances.

O que fazer com o ENTULHO????

1- Todo gerador de **ENTULHO** e responsável pela destino do mesmo. O entulho deve ser destinado ao Aterro de Goiânia ou usina de reciclagem, deve-se contratar uma empresa especializada e devidamente credenciada , para o acondicionamento do **ENTULHO** em caçambas ou containeres adequados, e transporte do mesmo ao local correto.

Observação : Nada de contratar carroceiros para dar fim ao entulho ,eles quase sempre depositam o **ENTULHO** em locais inadequados.

O que fazer com MOVEIS e ELETRODOMESTICOS INUTILIZADOS?

1- E proibido abandonar **MOVEIS VELHOS e ELETRODOMESTICOS** nas vias publicas, calcadas, ruas e praças, a COMURG oferece o serviço de **CATA TRECO**, onde o cidadão liga **(62) 3524-8555 (Tele- Atendimento da COMURG)** e solicita a coleta do material com hora agendada.

Figure 47 - COMURG guidelines on disposing of rubble, old furniture and unused household appliances. Source: COMURG website (2016).

With regard to old furniture and household appliances, it is easier for the population to dispose of them than it is for them to dispose of RCC. The residents of Jardim do Cerrado I to IV are part of a more vulnerable social class, where the head of the household often has a low level of education, is a single mother, works all day in central Goiânia (25 km away), depends on public transport and faces a series of other daily difficulties, such as a shortage of money.

A family of this profile will rarely hire a specialised and duly accredited company to package the rubble generated by building a "puxadinho" at the back of the house or changing the floor in the kitchen. The most this family could do would be to pay a cart driver to take the waste to the landfill, but even this is not the practice among residents, as shown in Figure 26, aggravated by the cart drivers' lack of responsibility for disposing of it anywhere. In view of this, it would be naïve to claim that these people dispose of their MSW on unoccupied plots only out of a lack of awareness or ignorance of the health risks that improperly disposed waste causes.

Solid waste is a health problem because it contributes to the proliferation of vectors and rodents, which cause various diseases such as infectious diarrhoea, amoebiasis, salmonellosis, helminths such as ascariasis, teniasis, trachoma and other parasites.

> Cockroaches, which land and live in solid waste where they find fermentable liquids, are of very relative sanitary importance in the transmission of gastro-intestinal diseases, through the mechanical transport of bacteria and parasites from filth to food and through the elimination of infected faeces (BRASIL, 2007, 231).

The sanitary landfill is still considered the most viable way to solve the problems faced by the inadequate disposal of municipal solid waste, because it follows the technical standards of the Brazilian Association of Technical Standards (ABNT), does not cause damage to public health and seeks methods to allocate waste in the smallest area, as well as doing so safely (CONDE et al. 2014).

d) Rainwater drainage

The lack of a rainwater drainage system is one of the main problems caused by urbanisation. With the retention of water on the surface of the ground, numerous consequences arise that directly interfere with the population's quality of life (TUCCI, 1995).

> The impacts on the population are mainly caused by the inappropriate occupation of urban space. These conditions generally occur due to the following actions: ■ as there is no restriction in the Urban Master Plan of almost all Brazilian cities on the subdivision of areas at risk of flooding, the sequence of years without flooding is reason enough for entrepreneurs to subdivide unsuitable areas; - invasion of riverside areas, which belong to the public authorities, by the low-income population; - occupation of medium-risk areas, which are hit less frequently, but when they are, suffer significant damage (TUCCI, 2002, p. 21).

In addition to waterlogging and flooding, the lack of urban drainage leads to the appearance of diseases such as leptospirosis, diarrhoea, typhoid fever and the proliferation of anopheline mosquitoes, which can spread malaria, among other diseases (BRASIL, 2007).

Rainwater drainage in the neighbourhood is another basic sanitation problem. As described above, Goiânia has serious problems with urban drainage, both in the centre and in the outskirts. Jardim do Cerrado, which was a "planned" neighbourhood, does not have an urban drainage infrastructure compatible with that of a planned neighbourhood. The streets in the neighbourhood have no curbs or manholes, and rainwater runs off at high speed, damaging the asphalt and forming erosions on the roads that are on the steeper, unpaved areas (Figures 48 and 49).

Figure 48 - Erosion caused by rainwater on Avenida Rainha dos Lagos in Jardim Cerrado III.
Source: Personal collection, Date: 17/12/2017.

There is no rainwater drainage system in several places in the neighbourhood. With the exception of a few streets, the neighbourhood has no kerbs or manholes. In other places, the tarmac has already been damaged by the force of rainwater (Figure 49).

Figure 49 - Lack of rainwater drainage infrastructure in Residencial Jardim do Cerrado I to IV.
Source: Personal collection. Date: 20/11/2016.

The lack of rainwater drainage also hampers residents' mobility in the neighbourhood due to the flooding and torrential downpours that form during the rainy season (Figure 50). In addition, various diseases such as leptospirosis are transmitted by the lack of rainwater drainage.

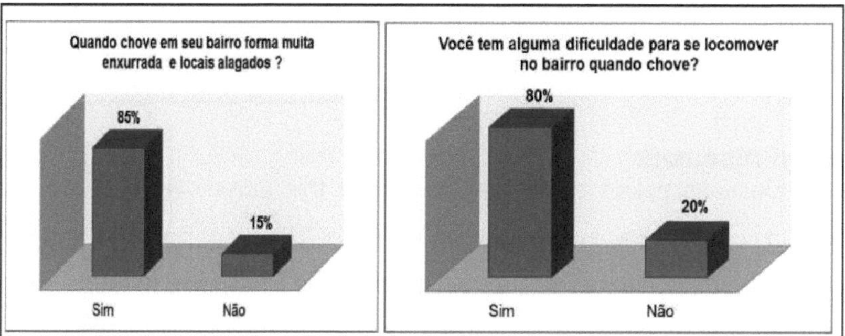

Figure 50 - Answers from Jardim Cerrado residents about rainwater drainage, Goiânia - GO.
Source: Prepared by the authors.

Residents of Jardim Cerrado III reported that until November 2016 there were no manholes on any street in the neighbourhood. But in the first week of December 2016 they were surprised by the rapid construction of some in a few streets in the neighbourhood. If you look at the bottom of these manholes, you can see that they are nothing more than a cemented hole inside with pre-moulded covers that are not connected to any rainwater drainage pipes (Figure 51).

Figure 51 - Manholes built irregularly in December 2016 in Jardim Cerrado III.
Source: Personal collection. Date: 17/12/2016.

The so-called planning that the real estate sector talks so much about boils down to neighbourhoods with water supply, sanitary sewage, asphalt, wide streets, public lighting, security (often related to walls and guardhouses), large *shopping centres* and supermarkets. As discussed in the first two chapters, urban planning is a complex issue that mainly involves the quality of life and well-being of the population. The lack of rainwater drainage can damage public roads, start erosion processes, cause flooding and consequently have a negative impact on public health. Tucci (2002) highlights some of the impacts on the population, namely: "material and human losses; interruption of economic activity in flooded areas; contamination by waterborne diseases such as leptospirosis, cholera, among others" (TUCCI, 2002, p. 11).

e) Acquired diseases

The last questions asked of the participants in this survey were about the illnesses they had suffered after living in Jardim Cerrado. The list of illnesses and the number of people who had already contracted any of them are shown in Table 16.

Table 16 - Main illnesses related to poor or non-existent basic sanitation affecting residents of Jardim Cerrado, Goiânia-GO.

PLACEMENT	DISEASE	N° PEOPLE
1°	Diarrhoea	51
2°	Dengue fever	46
3°	Skin diseases (mycosis)	30
4°	Amebiasis	24
5°	Conjunctivitis	20
6°	Hepatitis A	9
7°	Zika virus	8
8°	Yellow fever	4
9°	Leptospirosis	4
10°	Chikungunya fever	0
11°	Schistosomiasis	0
12°	Cholera	0

Source: Prepared by the authors.

Diarrhoea and dengue fever were the most common diseases among the participants in this survey. The risk factors associated with these diseases can be explained by various "socio-economic, political, demographic, health, environmental and cultural factors" (BENÍCIO et al, 1989). The main sanitation problems in the neighbourhood include solid waste, urban drainage and sewage.

Among the numerous urban problems related to poor municipal solid waste management are epidemics of dengue fever, lesptospirosis (FERREIRA, ANJOS, 2001), chikungunya, Zika, among others. Although there is regular domestic waste collection in the neighbourhood, the population still improperly disposes of many types of solid waste in vacant lots and public places. This encourages the proliferation of vectors and rodents that cause many of the diseases described in

Table 16.

Septic tanks need regular maintenance, otherwise their capacity to biodegrade effluent is exceeded and they end up becoming an environmental problem, attracting cockroaches, flies and other vectors that transmit diseases, especially parasites.

CHAPTER 5

FINAL CONSIDERATIONS

Urban and environmental planning in Brazil has gone through several phases, with countless laws regulating urban and environmental planning instruments in an attempt to promote a more orderly development of the country with a view to reducing regional inequalities. The Statute of the City and the creation of the Ministry of Cities ushered in a new phase in Brazilian urban planning. There have been significant advances in the post-Statute of the City planning period, especially with regard to investments and legislation on basic sanitation.

The National Sanitation Policy, instituted by Law No. 11.445 of 2007, brought new perspectives to the country's sanitation sector, defining guidelines and rules for planning, regulation, inspection, social control and, above all, the establishment of universalisation of basic sanitation services (drinking water supply, sewage disposal, urban cleaning and solid waste management and urban rainwater drainage and management) (BRASIL, 2007).

It can be seen that the evolution of basic sanitation services in Brazil is growing slowly and unevenly. Deficiencies in this service have a direct impact on the health of the population and public spending on hospitalisations in the SUS. It can be seen that the states belonging to the more developed regions, such as the south-east and south, have good levels of quality and service in basic sanitation services, while the opposite is true of the less developed regions, such as the north, north-east and centre-west. This confirms the urban planning model established in Brazil, which is marked by profound socio-economic inequalities between regions. Despite the progress made in the sector from the 1990s to the present day, universalisation of basic sanitation in Brazil is still a major challenge to be overcome.

Bringing the discussion to Goiânia, we can see from the data survey carried out in this research that the conditions of basic sanitation services in the capital of Goiás are not so unsatisfactory when the average for Brazilian municipalities is taken into account. However, this does not mean that the services provided are efficient and cover the entire population. It can be seen that sanitation services are still inefficient, mainly due to the rate of sewage collection and treatment and the lack of rainwater drainage networks in most of the municipality. The number of hospitalisations of people with diseases related to the lack or precariousness of basic sanitation is also indicative of the sanitary conditions of the population of Goiânia.

The number of hospitalisations for diseases related to the lack or inefficiency of basic sanitation shows that Goiânia needs improvements and greater coverage of the services offered. Of the total number of hospitalisations in the period assessed (2008-2016), the most vulnerable age group for most of the diseases studied were children aged between zero and four. Diarrhoea is one of the biggest child health problems in the world and one of the main causes of child mortality (BLACK etal., 2003).

The rate of water service in the urban area of Goiânia is already largely universalised and the rate of sewage collection is well above the national average. As far as sewage is concerned, one of the major challenges to achieving universalisation lies in the rural areas, which still have precarious sewage disposal solutions, and in the urban areas, where there is a deficit in both sewage collection and treatment.

When it comes to water supply, the rate of water loss in distribution must be taken into account. In this respect, it should be borne in mind that the basic sanitation service concessionaire in Goiânia has invested seriously in this area and the result was a loss of 21.07% in 2014, one of the four best rates in all of Brazil.

The ratio between investment and revenue in Goiânia is still low, i.e. only 25 per cent of all revenue obtained from operating income and through public authorities between 2010 and 2014 was invested. This low level of investment had a direct impact on the evolution of sanitation indicators, since there was no significant evolution in sewage treatment and total water service, which remained practically stagnant during this period. It is believed that if the proportion of investment increases and is applied more evenly across the various types of sanitation services, the coverage of all the indicators will grow concomitantly, making it possible to achieve universalisation of these services in the medium and long term.

The collection and environmentally appropriate disposal of solid urban waste in Goiânia has grown significantly in recent years and will continue to do so. With the Selective Collection Programme, the amount of recyclable materials no longer sent to landfill has been significant. The challenge to be overcome today is to find effective environmental education strategies in order to create environmental awareness among the population about the importance of not disposing of household or construction waste inappropriately in public places or vacant lots. In addition, for the selective collection programme to continue to work well and to be expanded, it is necessary for the population to support the process of separating recyclable waste at home and placing it in the bins on the scheduled days and times for collection by the selective collection truck.

Finally, the last and possibly most challenging sanitation deficit to be tackled refers to rainwater drainage because it is directly linked to the structure and construction process of the city. Every year, Goiânia faces flooding at different points, both in the central region and in noble and peripheral neighbourhoods. One of the most critical places is precisely in the central region, due to the steepness of the terrain and the high level of soil sealing. The solution to this problem involves breaking paradigms, financial investment, efficient short-, medium- and long-term government plans and, above all, the goodwill and interest of the various players involved in the city's administration.

It is important to highlight the situation of basic sanitation services in Jardim Cerrado, a new and "planned" neighbourhood that was established in 2009/2010, a year when various urban policy rules and urban and environmental planning instruments were already in force. The image of a planned neighbourhood, which is publicised in the media by the developer, highlights the distorted conception of planning that the real estate market transmits to society. Also evident in the neighbourhood is the unequal access to basic sanitation services faced by the less economically favoured population.

The social and physical inequality of Jardim Cerrado is visible. Modules I to IV, which were allocated to the most vulnerable families belonging to band 1 of the PMCMV, are marked by precarious buildings and infrastructure, including the lack of a sewage system, as the survey shows.

These and other inequalities in the infrastructure of the Jardim Cerrado residential modules make it clear that city planning is driven by the property market, which in turn seeks to maximise capital rather than the quality of life and well-being of the community.

Two hypotheses were defined at the beginning of this research, the first of which was that Goiânia still has basic sanitation deficiencies, especially in relation to sewage. Although Goiânia's sanitation indices are above the national average, the first hypothesis was confirmed at the end of this study. Goiânia still has many deficiencies in basic sanitation services, especially with regard to sewage and rainwater drainage. The precariousness or inefficiency of these services has a profound impact on the population's quality of life and health, as discussed in the fourth chapter.

The second hypothesis was: The number of cases of illness, hospital admissions and deaths caused by the lack or inefficiency of basic sanitation in Goiânia is still high, especially in the first years of life. The number of hospital admissions for diseases such as dengue fever, haemorrhagic fever, diarrhoea and other intestinal infectious diseases is still high in the municipality of Goiânia and mainly affects children between the ages of zero and four.

Further research should include a greater number of diseases related to the absence or inefficiency of basic sanitation and the collection of primary information through a population sample from peripheral and affluent neighbourhoods, in order to establish a comparison between socio-economic conditions and basic sanitation with the types of diseases acquired by each group studied.

REFERENCES

ABIKO, Alex Kenya; ALMEIDA, Marco Antonio Plácido de; BARREIROS, Mario Antonio Ferreira. Urbanism: History and Development. São Paulo: EPUSP, 1995.

ABRELPE. Panorama of Solid Waste in Brazil. Available at: http://www.abrelpe.org.br/Panorama/panorama2012.pdf. Accessed 14 Aug 2016.

ALBANO, M. P. (2013). The importance of urban environmental planning - social housing and urban expansion in Presidente Prudente-SP (Master's thesis).

Universidade do Oeste Paulista, Presidente Prudente.

ALMEIDA, J. R., MARQUES, T., MORAES, F. E. R., Bernardo, J. (1999). Environmental planning: the road to popular participation and environmental management for our common future - a necessity, a challenge (2nd ed.). Riode Janeiro: Thex Ed.

ALVES, D.; BELLUZZO W. Infant mortality and child health in Brazil, Economics & Human Biology, 2004.

AMORIM, Letícia Balsamão. On the claim for correction of unjust law. Revista de Informação Jurídica, a. 43, n. 171, p. 285-296, jul./set. 2006.

ANTONUCCI, A., Alvim, A. T. B., Zioni, S., & Kato, V. C. (2010). UN-Habitat: from declarations to commitments. São Paulo: Romano Guerra.

BALL, G.L. 1994 Ecosystem modelling with GIS. Environmental Management, 18(3): 345-349.

BENEVOLO, Leonardo. Diseno de la Ciudad - 5. El ambiente de la Revolución Industrial. 3. ed. Barcelona: Editora Gustavo Gilli S.A, 1982.

BLACK RE, MORRIS SS, BRYCE J. Where and why are 10 million children dying every year? Lancet 2003; 361: 2226-34.

BARROS, R. T. V. et al. Sanitation. Belo Horizonte: UFMG Engineering School, (Manual of sanitation and environmental protection for municipalities-volume 2) 1995.

BASSANI, P.; CARPIGIANI, P. H. C. Notes on the environmental movement and sustainable development. Analecta, Guarapuava, Paraná, v. 11, n. 1, p. 35-52, jan./jun. 2010.

BENICIO MHDA, CÉSAR CLG, GOUVEIA MC. Morbidity profile and pattern of utilisation of health services among Brazilian children under five years of age 1989. In: Monteiro MFC, Cervini R, organisers. Statistical profile of children and mothers in Brazil. Aspects of children's health and nutrition in Brazil 1989. Rio de Janeiro: IBGE, Unicef; 1992. p. 79-96.

BRAZIL. Constitution of the Federative Republic of Brazil, 1988.

BRAZIL. City Statute. Chamber of Deputies, 2ª Ed, 2001.

BRAZIL. Participatory Masterplan: guide for preparation by municipalities and citizens. Brasilia: Ministry of Cities; Confea, 2007.

BRUMES, Karla Rosário. Cities: (Re) Defining Their Roles Throughout History. Caminhos da Geografia, v. 2, n. 3, p. 47-56, mar. 2001.

CABANILLAS, Rolando Eli Quispe. Presentation - Strategic city planning: an alternative for the peoples of South America and the developing world. Campinas - São Paulo, Nov 2005. Website: www.cori.unicamp.br/ct/latinos-apres/seminariointernacional.pppt access: 15 November 2015.

CÂMARA, Jacintho Arruda. Masterplan. In: DALLARI, Adilson Abreu; FERRAZ, Sérgio (coord.) Estatuto da Cidade (comentários à Lei Federal 10.257/2001). 2. ed. São Paulo: Malheiros, 2006, p. 315-334.

CANEPA, C. (2007). Sustainable cities: the municipality as the locus of sustainability. São Paulo: RCS Editora. Brazil. Ministry of the Environment (1992). Agenda 21. United Nations Conference on Environment and Development (UNCED). Rio de Janeiro: Ministry of the Environment. Retrieved 04 July 2015, from http://www.mma.gov.br/responsabilidade- socioambiental/agenda-21 /agenda-21 - global.

CASTELLS, M. (1972). The urban question. Rio de Janeiro: Paz e Terra, 1983. 4ª Ed.

CAVENAGHI, Paula Talmelli. LIMA, Mariana. Plano Diretor: como a geotecnologia tem facilitado a gestão dos municípios. 2009 Site: www.img.com.br access: 15 November 2015.

CAVINATTO, V. M. Saneamento básico: fonte de saúde e bem- estar. São Paulo: Ed. Moderna, 1992.

COSTA, A. M. Evaluation of the national sanitation policy. 1996/2000. 248 f. Thesis (Doctorate in Public Health) - National School of Public Health, Oswaldo Cruz Foundation, Recife, 2003.

COSTA, H. S. M.; MENDONÇA, J. G. Novelties and permanence in the production of metropolitan space: a look at Belo Horizonte. In: OLIVEIRA, F. L.; COSTA, H. S. M.; CARDOSO, A. L.; VAINER, C. B. (Org). Major projects metropolitan: Rio de Janeiro and Belo Horizonte. Rio de Janeiro: Letra Capital, 2012.

COETZER PWW, KROUKAMP LM. Diarrhoeal disease - epidemiology and intervention. S Afr Med J 1989; 76: 465-72.

CYMBALISTA, R. A trajectória recente do planejamento territorial no Brasil: apostas e pontos a observar. Revista Paranaense de Desenvolvimento, n. 111, p. 29-45, Jul./Dec. 2006.

DA COSTA, A. V. F.; MAMEDE-NEVES, M. A. Photographic Images of Female Teachers: a visual trajectory of teaching in municipal schools in Rio de Janeiro in the late 19th and early 20th centuries. Rio de Janeiro, 2008, 243 p. PhD Thesis - Department of Education, Pontifical Catholic University of Rio de Janeiro.

DENÈGRE, J. 1994. Technological progress in geographical research: recent developments in satellite remote sensing and geographical information systems.

Mapping Sciences and Remote Sensing, 31 (1):3-12.

DI SARNO. D. C. L. Elementos de direito urbanístico. Barueri, SP: Manole, 2004.

DIÁRIO DE GOIÁS, 2011. Available at: http://www.blogdoandersonpereira.com/2011/05/moradores- denunciam-falta-de-saneamento.html. Accessed on: 13 August 2016.

DIÁRIO DE GOIÁS, Available at: 2014. http://diariodegoias.com.br/cidades/20463-tempestade-alaga-ruas- de-goiania-marginal-botafogo-transbordou. Accessed on: 13 Aug 2016.

DIAS, D. S. (2005). Urban development: constitutional principles. Curitiba: Juruá.

FELDMAN, S. São Paulo, 1947-1972: planning and zoning, PhD thesis presented to the Faculty of Architecture and Urbanism at USP, 1996.

FIDALGO, E.C.C. Criteria for analysing environmental methods and indicators used in the diagnosis stage of environmental planning. Doctoral Thesis - UNICAMP. Campinas, 2003.

FRANCO, M. A. R. (2001). Planejamento ambiental para acidade sustentável (2nd ed.). São Paulo: Annablume.

G1 da Globo, 2015. Available at: http://g1.globo.com/goias/noticia/2015/02/chuva-e-ventania- provocam-queda-de-arvores-em-varios-bairros-de-goiania.html. Accessed on: 13 Aug 2016.

GALVÃO JÚNIOR, A.C.; SOBRINHO, G.B.; SILVA, A.C. (2012) Panel of Indicators for Basic Sanitation Plans. In: PHILIPPI JÚNIOR, A. & GALVÃO JÚNIOR, A.C. (Ed.). Basic Sanitation Management: water supply and sewerage. Barueri: Manole. p. 1040-1068.

GOIÂNIA. Goiânia Statistical Yearbook 2012. Available at: http://www.goiania.go.gov.br/shtml/seplam/anuario2012/_html/d _bairros.html. Accessed on: 14 Aug 2016.

GOIÂNIA. Goiânia City Hall - COMURG 2008 Available at: http://www.goiania.go.gov.br/shtml/coletaseletiva/principal.shtml. Accessed on: 13 Aug 2016.

______. Goiânia City Hall - PUBLIC TENDER NOTICE No. 003/2015 Available at: https://www.goiania.go.gov.br/sistemas/silic/dados/m003/2015/L003 001201500030001.pdf. Accessed on 20 December 2016.

GROSTEIN, Marta Dora. Metropolis and Urban Expansion: the Persistence of "Unsustainable" Processes. São Paulo Perspect. v.15, n. 1, p. 13-19, jan./mar. 2001.

IANNI, O. *Estado e Planejamento Económico no Brasil.* 4ª ed. Rio de Janeiro: Civilização Brasileira, 1986.

HELLER, L. Sanitation and Health. Brasília: PAHO/WHO, Brasília 1997. COSTA, A.M.; PONTES, C.A.A.; GONÇALVES, F.R.; LUCENA, R.C.B.; CASTRO, C.C.L.; GALINDO, E.F.; MANSUR, M.C. (2010) Impacto na saúde e no Sistema Único de Saúde decorrentes de agravos relacionados a um sanamento ambiental inadequado. In: National Health Foundation. First research notebook on public health engineering. Brasília: National Health Foundation, p. 7-27.

HELLER. L. Institutional and Legal Framework of the Sanitation Sector in Brazil. IN. SIMPÓSIO LUSO-BRASILEIRO DE ENGENHARIA SANITÁRIA E AMBIENTAL, VIL, 1996, Lisbon. Proceedings. Lisbon: ABES/APRH, 1996. p.32-43.

HELLER, L. *Sanitation and health.* Brasilia: PAHO, 1997.

HELLER, L.; NASCIMENTO, N.O. Research and development in sanitation in Brazil: needs and trends. Eng Sanit Ambient. 2005; 10(1): 24-35.

HENDRIX, w.g.; FABOS, J.G. & PRICE, J.E. 1988. An ecological approach to landscape planning using geographic Information system technology. Landscape and Urban Planning, 15:211-225.

BRAZILIAN INSTITUTE OF GEOGRAPHY AND STATISTICS. 1991 Demographic Census. Available at:http://www.ibge.gov.br/home/estatistica/populacao/censodem/default_censo1991.shtm. Accessed on: 14 July 2016.

. Demographic Census 2000. Available at: http://www.ibge.gov.br/home/estatistica/populacao/censo2000/. Accessed on: 14 July 2016.

. Demographic Census 2010. Available at: http://censo2010.ibge.gov.br/. Accessed on: 14 July 2016.

JORGE, Karina Camarneiro. Urbanism in Empire Brazil: Public Health in the City of São Paulo in the 19th Century (Hospitals, Lazarets and Cemeteries). 2006. 226 f. Dissertation (Master's in Urbanism) - Pontifical Catholic University of Campinas - Centre for Exact, Environmental and Technological Sciences. Campinas, 2006.

JORNAL REGIONAL, The presence of faecal coliforms in water, Brasília, 2009 Available at at http://www.jornalregional.com.br/noticia/1281>. Accessed 10 August 2016.

KOHLSDORF, M. E. (1985). A brief history of urban space as a disciplinary field. In O espaço da cidade - contribuição à análise urbana (pp. 15-72). São Paulo: Projeto.

LAFER, Betty Mindlin. Planning Brazil. São Paulo, Ed. Perspectiva, 1973.

LEE, N.; WOOD, C.M. & GAZIDELLIS, V. (ed) 1985. Arrangements for environmental impact assessment and their training implications in the European Communities and North America: country studies. Occasional Paper n.13. University of Manchester,

Dep. of Town and Country Planning. Manchester. 311p.

LEES, B.G. RITMAN, K. 1991. Decision-tree and rule-induction approach to integration of remotely sensed and GIS data in mapping vegetation in disturbed or hilly environments. Environmental Management, 15(6): 823-831.

LIMA NETO, LE. & SANTOS, A.B.D. (2012) Basic Sanitation Plans. In: PHILIPPI JÚNIOR, A.; GALVÃO JÚNIOR, A.C. (Org.). Gestão do Saneamento Básico: abastecimento de água e esgotamento sanitário. Barueri: Manole. p. 57-79.

LISBOA, S. S; HELLER, L; SILVEIRA, R. B, Challenges of municipal sanitation planning in small municipalities: the perception of managers, Eng. Sanit Ambient | v.18 n.4 | oct/dec 2013 | 341-348.

LUIZ, G. C. Influence of the soil-atmosphere relationship on the hydromechanical behaviour of unsaturated tropical soils: case study - municipality of Goiânia GO. (DOCTORATE THESIS) University of Brasilia Faculty of Technology Department of Civil and Environmental Engineering, 2012.

MACEDO, H. Saneamento e saúde - um estudo de caso da vila Roriz, em Goiânia/Goiás, (Master's dissertation) postgraduate programme in geography, University of Brasília - UNB, 2008.

MADEIRA, R. F. The basic sanitation sector in Brazil and the implications of the regulatory framework for universal access. Revista do BNDES 33, p. 123-154, June 2010.

MARIA, Y. R. (2013). Urban solid waste and public environmental education policies: the case of Pontal do Paranapanema-SP (Master's thesis). Universidade do Oeste Paulista, Presidente Prudente.

MARICATO, E. Brasil, cidades: alternativas para a crise urbana. Petrópolis: Vozes, 2008.

METZGER, J. P. Is the Forest Code Scientifically Based? Conservation and Nature, v. 8, n. 1, p. 92-99, 2010.

MINISTÉRIO DA SAÚDE, PORTAL DA SAÚDE SUS, 2015 - Vaccination.Available at: http://portalsaude.saude.gov.br/index.php/o-ministerio/principal/leia- mais-o-ministerio/631-secretaria-svs/vigilancia-de-a-z/colera/l2- colera/11175-vacinacao-colera. Accessed on: 31 January 2017.

. PORTAL DA SAÚDE SUS, 2014. Available at: http://portalsaude.saude.gov.br/index.php/o-ministerio/principal/leia- mais-o ministerio/691 -secretaria-svs/vigilancia-de-a-z/febre- tifoide/U-febre-tifoide/13930-febre-tifoide. Accessed on: 31 January 2017.

MONTEIRO, Circe Maria Gama. Planning: Some Considerations. Espaço, Tempo e

Crítica, v. 1, n. 1(2), p.40-54, 2006.

MORAES, I. R. The urbanisation process and the Neighbourhood Impact Study - EIV. IN: CONPEDI, 2007, Belo Horizonte Desafios do Direito Urbanístico, 2007.

MORAES, L, M.. Planned Integration: Goiânia, Brasília and Palmas. Goiânia: Ed. da UCG, 2003.

MORAES LRS. Evaluation of the impact on health of environmental sanitation actions in impoverished areas of Salvador - AISAM Project. In: Heller L, Moraes LRS, Monteiro TCN, Salles MJ, Almeida LM, Câncio J, organisers. Sanitation and health in developing countries. Rio de Janeiro: CC& P; 1997. p. 281305.

MOTTA, A. P. C.; BAUMAN, C. M.; BRUNES, R. R. População em situação de rua: contextualização e caracterização, Revista Virtual Textos & Contextos, n° 4, dez. 1994.

PEGARORO, D. B. The institutional implementation of the Neighbourhood Impact Study and the consolidated practices of other impact studies. Master's dissertation. Postgraduate Programme in Urban and Regional Planning, Federal University of Rio Grande do Sul, 2010.

PHILIPPI, A., Jr., ROMÉRO, M. A., & BRUNA, G. C. (2004). An Introduction to the Environmental Question. In: A. Philippi Jr., M. A. Roméro, & G. C. Bruna (Orgs.). Environmental Management Course (p. 3-16). Barueri: Manole.

PINSKY, Jaime. The First Civilisations. São Paulo: Contexto, 2001.

PLANSAB - National Plan for Basic Sanitation, more health with quality of life and citizenship, Brasilia, May 2013.

RIBEIRO, P.J. (2004) Educational programme on schistosomiasis: a methodological approach model. Revista de Saúde Pública, v. 38, n. 3, p. 415-421.

RIGGS, J. L. Aids Transmission in Drinking water: no threat. Journal of amrican water works association, v. 81, n. 9, p. 69, 1989.

RODRIGUES, L. P. O. S.; FILHO, N. B.B. O controle das atividades urbanas e as mudanças climáticas: enfoque sobre a futura região metropolitana de São Luís do Maranhão. In: Caderno de Pesquisas, Universidade Federal do Maranhão, São Luís, v. 18, n. 1, jan/abr. 2011, 14p. Available at:< http://pppg.ufma.br/cadernodepesquisa/upload/files/Artigo%203%28 22%29.pdf>. Accessed on: 22 December 2015.

RODRIGUEZ, J. M. Apuntes de geografia de los paisajes. Havana: Imprenta Andre Voisin, 1991.

ROLNIK, R. (2001). City Statute - an instrument for cities that dream of growing in justice and beauty. In: N. Saule Jr., & R. Rolnik (Eds.), City Statute: new horizons for urban reform (Caderno Pólis, n. 4, p. 5-9). São Paulo: Pólis. Retrieved 06 February 2016, from www.polis.org. br/obras/arquivo_92. pdf.

ROMEIRO, A. R. Sustainable development: an economic-ecological perspective. Estudos Avançados, v. 26 n. 74, 2012.

ROSEN, George: A History of Public Health. São Paulo: Ed. UNESP, 1994. p. 223.

SANTOS dos, R. F., Construction of scenarios in a GIS environment to evaluate land use changes induced by hydroelectric power plants in the agricultural region of Andradina (SP). Master's dissertation - UNICAMP. Campinas, 2013.

SANTOS, M. a natureza do espaço: técnica e tempo, razão e emoção. São Paulo: Hucitec 1998.

SAULE JÚNIOR, Nelson. The legal protection of housing in irregular settlements. Porto Alegre: Sérgio Antônio Fabris Editor, 2004.

SCHVASBERG, B. Urban planning in post-1988 Brazil: historical panorama and contemporary challenges. In: Seminar on Urban and Regional Policies in Brazil. UnB - Faculty of Architecture and Urbanism. Nov. 2009.

SILVA FILHO, A. O; RAMOS, J. M; OLIVEIRA, K; NASCIMENTO, T. The Evolution of the Brazilian Forest Code. Human and Social Sciences Unit, Aracaju, v. 2, n.3, p. 271-290, March 2015.

SILVA, João dos Santos.Vila. Multivariate analysis in zoning for environmental planning; case study: upper Taquari river basin MS/MT. PhD Thesis - UNICAMP - Campinas, 2003.

SILVA,R. T.; MACHADO, L. Networked urban services and public control of the subsoil - new challenges for urban management. Sao Paulo Perspect. 2001; 15(1): 102-11.

SILVESTRE, M. E. D. 1934 Code: Water for Industrial Brazil. Rev. Geo-Paisagem, ano 7, n 13, jan./jun. 2008.

SELLTIZ, C.; JAHODA, M.; DEUTSCH, M.; COOK, S. W. Research Methods in Social Relations. Revised edition. Translation by Dante Moreira Leite. 5ª reprint. São Paulo: E.U.P., 1975.

SILVA, C. H. R. T. Stockholm '72, Rio de Janeiro '92 and Johannesburg '02: the three major international environmental conferences. Legislative Bulletin, n. 6, 2011.

SILVA, P. D. O. D'; LOLLO; J. A. The neighbourhood impact study as an instrument for the development of urban quality of life. Holos Environment, v.13, n.2, 2013, p. 151.

SILVEIRA, A.C. (2013) State of vector transmission control of Chagas disease in the

Americas. Cadernos de Saúde Pública, v. 16, sup. 2, p. 35-42.

SIQUEIRA, D. J. Environmental Impact Assessment - EIA/RIMA. (Handout). Londrina: Federation of Engineering and Architecture Associations of Paraná /CREA PR, 2002.

SISAGUA and Ministry of Health, National Guideline for the Environmental Health Surveillance Sampling Plan related to the quality of water for human consumption. 2ª ed. 2013 . Available at:< http://www.saude.gov.br/bvs> Accessed on: 13 January 2016.

SNIS - Sistema Nacional de Informações Sobre Saneamento, Manual de preenchimento dos prestadores de abrangência local de água e esgotos, 2014. Available at:< http://www.snis.gov.br/coleta-de-agua-e-esgotos> Accessed on: 22 March 2016.

SOARES, S. R. A.; BERNARDES, R. S.; CORDEIRO NETTO, O. M. Relationships between sanitation, public health and the environment: elements for formulating a sanitation planning model, Cad. Saúde Pública, Rio de Janeiro, v 18 n 6, p. 17131724, 2002.

SOUZA, Marcelo Lopes de. Mudar a Cidade: Uma introdução Crítica ao Planejamento e à Gestão Urbanos. Rio de Janeiro: Bertrand Brasil, 2008.

SOUZA, Marcelo Lopez de. Mudar a cidade: uma introdução crítica ao planejamento e gestão urbana. 5 edition, Rio de Janeiro - RJ: Bertrand Brasil, 2008a.

TV GUAIAMUM, 2016, Available at: http://www.tvguaiamum.com/2016/03/populacao-da-portelinha- reclama-de.html. Accessed on: 13 Aug 2016.

UNICEF and WHO, World Health Organisation says too few have access to improved sanitation, 2015, Available at:< http://www.unicef.org/brazil/pt/media_12597.htm> Accessed on: 30 March 2016.

VALLADARES, L. P. do, A gênse da favela carioca: a produção anterior às ciências sociais. In: Revista Brasileira de Ciências Sociais, v. 15, n. 44. Oct. 2002.

VIESSMAN Jr., W. & SCHILLING, K. (eds) 1986. Social and environmental objectives in water resources planning and management. American Society of Civil Engineers. New York, 325p.

VIGIL, Percy Acuna. The City in the Modern Age. Lima (Peru): UNI/FAUA, 2003.

VILLAÇA A contribution to the history of urban planning in Brazil. In: DÉAK C.; SCHIFFER, S. R. (Org.). O Processo de urbanização no Brasil. São Paulo: FUPAM/EDUSP, 1999. p. 169244.

VILLAÇA, F. Espaço interurbano no Brasil. São Paulo: Studio Nobel, 1998.

VILLAÇA. Perspectives on urban planning in Brazil today. In: II Seminar on Brazilian Cities - Desires and Possibilities, organised by the Campo Grande City Council, MS. Campo Grande. 2000. Available at at:<http://www.flaviovillaca.arq.br/pdf/campo_gde.pdf>. Accessed on: 13 Apr. 2016.

ABOUT THE AUTHORS

Lúcia Maria Moraes

Architecture and Urban Planning. PhD in Urban Environmental Structures from the University of São Paulo.
She is currently a lecturer and advisor on the Master's Programme in Social Work and the Master's Programme in Development and Territorial Planning, as well as on the following courses
Architecture and Urbanism, Pontifical Catholic University of Goiás - PUC Goiás.
lucia.dhescmoradia@gmail.com

Obede Rodrigues Alves

Environmental Engineer. Master's in Territorial Development and Planning from Pontifical Catholic University of Goiás - PUC Goiás.
PhD student in Environmental Engineering Sciences - Centre for Water Resources and Environmental Studies, São Carlos School of Engineering, University of São Paulo, CRHEA/EESC/USP. São
Paulo SP, São Carlos.
alves.obede@gmail.com

Otniel Alencar Bandeira

Nurse. Specialist in Public Health, Family Health and Labour Nursing.
Master's degree in Territorial Development and Planning from the Pontifical Catholic University of Goiás (PUC Goiás).
otnielalencar@gmail.com

Printed by Books on Demand GmbH, Norderstedt / Germany